# Wind Energy in the 21st Century

*Also by Robert Y. Redlinger*

TOOLS AND METHODS FOR INTEGRATED RESOURCE PLANNING
(*with Joel Swisher and Gilberto Jannuzzi*)

# Wind Energy in the 21st Century

## Economics, Policy, Technology and the Changing Electricity Industry

Robert Y. Redlinger
*Regional Manager, Distributed Generation*
*CMS Viron Energy Services, California*

Per Dannemand Andersen
*Senior Scientist*
*Risø National Laboratory, Denmark*

and

Poul Erik Morthorst
*Senior Research Specialist*
*Risø National Laboratory, Denmark*

palgrave

UNEP
United Nations Environment Programme

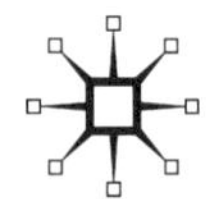



First published 2002 by
PALGRAVE
Houndmills, Basingstoke, Hampshire RG21 6XS and
175 Fifth Avenue, New York, N. Y. 10010
Companies and representatives throughout the world

PALGRAVE is the new global academic imprint of
St. Martin's Press LLC Scholarly and Reference Division and
Palgrave Publishers Ltd (formerly Macmillan Press Ltd).

ISBN 0–333–79248–3

This book is printed on paper suitable for recycling and made from fully managed and sustained forest sources.

A catalogue record for this book is available from the British Library.

Library of Congress Cataloging-in-Publication Data
Redlinger, Robert Y., 1963–
Wind energy in the 21st century : economics, policy, technology, and the changing electricity industry / Robert Y. Redlinger, Per Dannemand Andersen, and Poul Erik Morthorst.
p. cm.
Includes bibliographical references and index.
ISBN 0–333–79248–3
1. Wind power. I. Title: Wind energy in the twenty–first century. II. Andersen, Per Dannemand, 1958– III. Morthorst, Poul Erik, 1951– IV. Title.

TJ820.R43 2001
333.9'2—dc21

2001021717

10 9 8 7 6 5 4 3 2 1
11 10 09 08 07 06 05 04 03 02

Printed and bound in Great Britain by
Antony Rowe Ltd, Chippenham, Wiltshire

# Contents

# List of Figures

# List of Tables

# Foreword: Signposts to Sustainability

## Klaus Toepfer

There are few signposts on the path to sustainable development more visible – and more inspiring – than the rapid evolution of the modern wind energy industry. In just two short decades, from 1980 to the year 2000, the industry grew from a few experimental turbines to a world market worth several billion dollars annually and an installed capacity of over 13 000 megawatts. These figures are not just impressive, they are several times what was considered to be even a highly optimistic scenario in the early 1990s.

The modern wind energy industry is a successful example of what can be achieved when governments combine the right investment signals with adequate support for research and development. Although the development model may not be the same for other sustainable energy technologies, the lessons from wind are timely and useful.

At the beginning of a new millennium, there are great hopes for wind energy to provide a significant portion of the electricity needed to serve a population that is expected to reach 9 billion people before the first five decades of this century are over. This type of growth for wind, considered an unachievable dream just a few years ago, is now a solid and achievable goal – but only if wind can continue its rapid development path. This will require both technical and policy improvements.

On the technical side, wind energy developers are well placed to reach further cost reductions through a rapid 'learning curve' inherent in building and operating thousand of turbines. However, wind developers will be challenged at the same time to ensure that their industry is sustainable through the production and use of turbines that are fully recyclable and socially acceptable, as well as manufacturing processes that are non-toxic and based on renewable materials.

On the policy side, governments must continue to remove subsidies that encourage fossil fuels and inefficient energy use and to ensure that prices for electricity tell the environmental truth. This will be particularly important in new and evolving competitive markets for electricity.

*Wind Energy in the 21st Century: Economics, Policy, Technology and the Changing Electricity Industry* is an important contribution to achieving these goals. The book provides a comprehensive analysis of the technical, economic and social dimension of the modern wind energy industry, including the global potential of wind energy technologies, the wind resource potential, scenarios for future development, and the economic and social impacts of wind energy development. Wind developers, government officials and other stakeholders will be able to use this information to develop policies and strategies to increase the development and application of wind energy technologies.

The book is the result of substantial co-operation between the United Nations Environment Programme (UNEP), the Department of Economic and Social Affairs (DESA) and the UNEP Collaborating Centre on Energy and Environment (UCCEE) at Risø National Laboratory in Denmark. UCCEE was established by UNEP as a global centre of excellence on sustainable energy issues and supports the work of UNEP's Energy programme. UCCEE is also a collaborative effort with the Danish Ministry of Foreign Affairs, who have generously supported the preparation and publication of this book.

I would like to congratulate all those who contributed to this outstanding work and I am confident that it will become a basic reference for anyone concerned with wind energy in the twenty-first century.

*Executive Director*
*United Nations Environment Programme*

# Preface

The beginning of the twenty-first century is an exciting time for wind energy. With the changes in technology, policy, environmental concern and electricity industry structure which have occurred in recent years, the coming decade offers an unparalleled opportunity for wind energy to emerge as a viable mainstream electricity source and a key component of the world's environmentally sustainable development path. Yet the challenges facing wind energy remain both substantial and complex.

This book resulted from the recognition that, for wind energy to continue its success and worldwide growth, policy makers and industry practitioners must understand the complex interplay of factors affecting wind energy today. These factors include not just technology and economics, but also issues of policy, finance, competition and environment.

We have written this book with three primary audiences in mind: policy makers, academic researchers and electricity industry professionals. For energy policy makers and energy industry analysts, the book should be useful for understanding how the wind energy industry has arrived at its current state, what have been the factors for success as well as failure, and, most importantly, what are the critical factors which will affect its future evolution and implementation.

Regarding academia, environmental policy issues are of significant interest to researchers and students, and the range of environmentally oriented courses has grown substantially over the past two decades. No substantial discussion of issues like climate change can take place without considering energy and its impact on the environment. This book will provide researchers and students with a broad understanding of not only the significant policy issues facing renewable energy, but also the critical impact of factors such as finance and electricity industry structure on wind power's viability.

For investors, interest in renewable energy is increasing due to a combination of factors including (a) the emergence of new markets such as the green electricity market; (b) the emergence of climate change as a potentially monumental force pushing the entire world's energy industry towards greater environmental sustainability; and (c) increasing levels of investment support becoming available for environmentally sound technologies through organisations such as the Global Environment Facility, the World Bank and bilateral donors. For investors contemplating an entry into the wind energy market, this book will provide a valuable understanding of the forces affecting the industry and its prospects for profitability.

This book has benefited from the input and assistance of many individuals and organisations. We would like to thank Dr S. Arungu Olende of the United Nations Department of Economic and Social Affairs who made this book possible and under whose direction the idea for the book was first conceived. We gratefully acknowledge the financial support of the Danish Ministry of Foreign Affairs who funded our work, and Dr John Christensen of the UNEP Collaborating Centre on Energy and Environment at Risø National Laboratory in Denmark for his invaluable support and guidance throughout the process.

We wish to thank the entire staff of the Wind Energy and Atmospheric Physics Department at Risø National Laboratory for their assistance with numerous parts of this book. We owe special thanks to Lars Landberg, Niels Gylling Mortensen and Erik Lundtang Petersen for their help on wind resource assessment, and Sten Frandsen, Helge Aagaard Madsen, Henrik Bindner and Poul Sørensen for their help with wind energy technology issues. We would also like to thank our colleagues at the UNEP Collaborating Centre on Energy and Environment and the Systems Analysis Department at Risø National Laboratory for their support and input, and, in particular, Kai Holst Andersen for his thoughtful and indispensable assistance throughout the publication process.

We are indebted to many other individuals who have provided valuable insights, expertise, and review comments regarding various parts of this book. They include (in alphabetical order): Pramod Deo of the Maharashtra State Electricity Board, Chris Ann Dickerson of Pacific Gas and Electric Company, Gaynor Hartnell of the British Wind Energy Association, Karen Marshall of Offer, Catherine Mitchell of the University of Sussex, Brian O'Gallachoir of the Irish Renewable Energy Information Office, Nancy Rader of the American Wind Energy Association, Ian Rowlands of the University of Waterloo, William Short of Ridgewood Power Corporation,

R. Suresh of the Tata Energy Research Institute, Eric Thompson of the Environmental Defense Fund, Maj Dang Trong of the Danish Energy Agency, Willem van Zanten of Novem Netherlands, Ryan Wiser of Lawrence Berkeley National Laboratory, and Maarten Wolsink of the University of Amsterdam.

Notwithstanding these acknowledgements, any errors contained in this book remain strictly the responsibility of the authors. Moreover, while the book was written under the auspices of the United Nations Environment Programme (UNEP), the United Nations Committee on New and Renewable Sources of Energy and on Energy for Development, and through funding from the Danish Ministry of Foreign Affairs, the views expressed herein are those of the authors and do not necessarily reflect the views of UNEP, the United Nations or the Danish Government.

R. Y. R.
P. D. A.
P. E. M.

# List of Abbreviations

| | |
|---|---|
| ACRS | accelerated cost recovery system |
| AER | alternative energy requirement |
| BEM | blade element momentum |
| BOOT | build, own, operate and transfer |
| CAA | computational aeroacoustics |
| CAPM | capital asset pricing model |
| CEC | California Energy Commission |
| CFB | circulating fluidised bed |
| CHP | combined heat and power |
| COADS | Comprehensive Ocean-Atmosphere Data Set |
| CPUC | California Public Utilities Commission |
| CV | contingent valuation |
| DESA | (UN) Department of Economic and Social Affairs |
| DKK | Danish kroner |
| DMI | Danish Meteorological Institute |
| DSCR | Debt-service coverage ratio |
| ECU | European Currency Unit |
| EFL | Electricity Feed Law |
| ERP | European Recovery Programme |
| FERC | (US) Federal Energy Regulatory Commission |
| GHG | greenhouse gas |
| HIRLAM | high resolution limited area model |
| IEA | International Energy Agency |
| IEC | International Electrotechnical Commission |
| IGCC | integrated gasification combined cycle |
| IPP | independent power producer |
| IREDA | Indian Renewable Energy Development Agency |
| ITC | investment tax credit |
| IWLA | Izaak Walton League of America |
| LEDC | local energy distribution company |
| LOLP | loss of load probability |
| LRMC | long-run marginal cost |
| MNES | Ministry of Non-conventional Energy Sources (India) |
| MV | megavolt |

| | |
|---|---|
| NFFO | (UK) Non Fossil Fuel Obligation |
| NGC | National Grid Company |
| NLG | Netherlands guilders |
| NPV | net present value |
| NSP | Northern States Power |
| NUG | non-utility generators |
| NWP | numerical weather prediction |
| O&M | operation and maintenance |
| PCC | point of common coupling |
| PTC | production tax credit |
| PURPA | (US) Public Utility Regulatory Policies Act |
| PV | photovoltaics |
| QF | qualifying facilities |
| RD&D | research, development and demonstration |
| RECs | renewable energy credits (USA) |
| | regional electricity companies (UK) |
| REFIT | Renewable Energy Feed-In Tariff (Germany) |
| RPS | Renewables Portfolio Standard (US) |
| SEBs | state electricity boards |
| SEK | Swedish kronor |
| SO | Standard Offers |
| UNFCCC | UN Framework Convention on Climate Change |
| UNEP | United Nations Environment Programme |
| USEIA | US Energy Information Administration |
| VAWT | vertical-axis wind turbines |
| VSL | value of a statistical life |
| WACC | weighted average cost of capital |
| WAsP | Wind Atlas Analysis and Application Program |
| WEC | World Energy Council |

*Dedicated to the memory of John K. Turkson, 1954–2000*

# 1
# Introduction

Renewable energy from the sun, wind and sea has long been touted as the ultimate solution to the world's energy and environmental problems, offering the potential of virtually unlimited cheap and pollution-free energy. Initial interest in renewable energy, spurred by the oil crises of the 1970s and fears of resource depletion and political insecurity, resulted in frenetic research and development activity, impressive technological advances and bold energy policy experiments. Yet, as the 1980s passed into the 1990s, fears of energy crises faded into the past and fossil fuel prices dropped to their lowest levels ever, while renewable energy technologies remained expensive in spite of the advances made. Renewables looked like they might forever remain 'the energy of tomorrow'.

However, even as renewable energy's overall fortunes declined, development of one particular renewable energy technology, the wind turbine, continued to make steady progress. With continuous improvements in reliability, efficiency and reductions in capital cost, the per kWh cost of wind energy declined by approximately 8 per cent per year throughout the 1990s. As a result, wind energy prices in the late 1990s have become roughly competitive in many cases against conventional forms of electricity generation.

At the same time, a new element of scientific and political uncertainty has once again entered the world energy picture: climate change. Following the Rio de Janeiro Earth Summit in 1992, political pressure has been increasing to address the growing concern over potential global climate change resulting from anthropogenic emissions of 'greenhouse gases' such as carbon dioxide. This process has led to the successful negotiation of the Kyoto Protocol in December 1997, committing industrialised nations to binding reductions in emissions of gases responsible for warming the atmosphere. With electric power generation accounting for a major share of greenhouse gas emissions, a new impetus has thus been created for increased implementation of renewable energy.

The third global trend which has emerged over the past decade has been the move towards privatisation, non-utility electricity generation and competition. And with this has come the opportunity for niche players such as wind power plants to enter the electricity market and, in some cases, provide specialised high value-added services such as sales to the 'green' electricity market. At the same time, the advent of competition has created new power market structures such as short-term forward and spot markets, creating added complexity for conventional and renewable generators alike.

These three large-scale trends of technological advance, global political change and electricity industry restructuring suggest that it is time to re-examine the status and prospects for wind energy. With an average annual growth rate in installed generation capacity of over 22 per cent per year between 1991 and 1997 (IEA, 1998) and estimated at over 35 per cent in 1998 (AWEA, 1999), wind energy is currently not only the most promising renewable energy technology, but also the world's most rapidly growing energy source, whether conventional or renewable. Over the next ten years, wind energy could complete its transition from 'alternative' to fully competitive 'mainstream' energy source. Whether this transition in fact occurs will depend on a complex interplay of factors including technological development, economics, finance, environment and policy. Policy makers as well as investors must understand these complex interactions in making appropriate decisions about the electricity industry in general and wind energy in particular.

This book aims to provide a thorough analysis of wind energy's current status, its future prognosis and the factors which will impact on its evolution over the coming decade. The book is divided into eight chapters, with separate chapters devoted to wind energy resource potential, technology and industry, economics, finance and power markets, environment and policy. The book is not meant to provide detailed technical analyses of topics such as wind turbine technology, wind speed measurement, or design guidance; such technical subjects are well addressed in the existing literature. Rather, this book provides, in one concise volume, sufficient coverage of a broad range of wind energy issues such that the policy maker, researcher or electricity industry professional can obtain a clear understanding of the most important issues facing wind energy as it enters the energy mainstream. Particular emphasis is placed on policy mechanisms to facilitate wind energy implementation, as well as the emerging issue of competitive power markets.

Different countries have used widely varying mechanisms to promote wind energy and have also had widely varying levels of success. This book provides case studies of the countries who have led the world in renewable energy development in general and wind energy development in particular, including the USA, UK, Netherlands, Denmark, Germany, India and Sweden. Their policy successes as well as failures are analysed and policy recommendations are provided.

The book is organised into the following chapters.

Chapter 2: 'Wind Energy Resource Potential'. Serious consideration of wind energy is only justified if the available wind resource potential is sufficient to meet a sizeable portion of countries' electricity needs. Furthermore, it is important to have a basic understanding of where wind energy currently stands in terms of installed capacity and countries' installation trends. Chapter 2 discusses worldwide wind resource potential, current installed capacity by country, industry forecasts of future installed wind capacity and selected governments' visions or targets for future wind energy growth. The chapter also introduces issues related to wind resource assessment.

Chapter 3: 'Wind Turbine Technology and Industry'. While wind energy has received the attention of the modern electricity industry for only the past two decades, wind energy utilisation in fact dates back hundreds and even thousands of years. Chapter 3 describes the development of wind energy technology as well as the basic principles for extracting energy from the wind. The chapter describes both wind technology and the wind industry and their likely future evolution. Wind turbines should not be examined solely in isolation but also in relation to the rest of the electricity grid. As wind turbine power output varies instantaneously in accordance with wind speed, such fluctuations can affect power quality throughout the electricity grid as well as the operational and dispatch demands placed on other power plants. Such grid-impact considerations are also addressed in the chapter, including wind energy's capacity credit, short-term wind prediction and power quality.

Chapter 4: 'Economics of Wind Energy'. Some of the most significant advances in wind energy have been in the area of cost reduction. Chapter 4 details the evolution of wind energy costs and provides a detailed breakdown of capital costs as well as operation and maintenance costs. Another new development in wind energy is the implementation of offshore wind turbines, of which three such wind farms currently exist in the world. Offshore wind energy costs are also analysed in detail in the chapter. A comparison between the economics of wind energy and conventional fossil fuel-based electricity is also provided. The emphasis of the chapter and of the book in general is on large-scale grid-connected wind electricity, which has dominated the development of wind energy in recent decades. However, the economics of smaller off-grid and hybrid applications are also addressed in less detail in Chapter 4.

Chapter 5: 'Finance, Competition and Power Markets'. Issues of finance can be quite distinct from those of economics, though the two are clearly related. Even when a wind energy facility appears to be economically cost-effective, the facility may often still not be built due to constraints in financing. Chapter 5 therefore addresses issues of wind power finance and the factors which must be accounted for in raising finance for wind power plants, such as risk, capital structure and output variability. Special issues of financing plants in developing countries are also addressed. The other major focus of Chapter 5 is the issue of competitive power markets which are emerging as part of a worldwide trend away from centralised vertically integrated monopoly utilities. Special challenges facing wind energy in competitive markets are discussed, including those regarding bidding into short-term forward and spot markets and transmission-related issues. Within the context of competition, the chapter also touches on opportunities in specialised niche markets like the 'green' market for 'environmentally friendly' electricity.

Chapter 6: 'Environmental Considerations'. Wind energy offers significant environmental benefits in terms of reduced air, water and ground pollution. However, wind energy can also have both real and perceived environmental drawbacks, including visual intrusion, noise, bird kills and others. In some locations, the local aesthetic impact of wind turbines may carry more weight than global environmental benefits such as reduced climate change. Thus, wind turbine projects often struggle for planning permission in spite of their credentials as some of the most environmentally benign energy sources. Chapter 6 addresses such environmental issues surrounding wind energy. It introduces the concepts of environmental amenity valuation and provides some comparisons between estimates of wind turbines' and other technologies' environmental impacts.

Chapter 7: 'Wind Energy Policy'. Though great advances have taken place in wind energy technology and economics, the impetus making such advances possible was, in large part, careful and thoughtful energy policy. The policies pursued by countries such as the USA, the UK and Denmark differ significantly and their levels of success similarly differ. Often, the policy environment is the single most important determinant of whether wind energy succeeds as a viable energy source. Chapter 7 therefore discusses the various policy mechanisms available for stimulating wind energy, including fixed power purchase contracts, production subsidies, tax credits, market set-asides, environmental taxation, preferential finance and so on. Case studies of the world's leading wind energy countries are provided, including an analysis of their successes and failures.

Chapter 8: 'Summary and Conclusions'. The final chapter summarises the earlier chapters and draws conclusions on the status of wind energy and its prospects over the coming decade. The basic questions addressed in this chapter include 'what are the main barriers preventing more widespread adoption of wind energy?' and 'what are the critical next steps necessary to enhance wind energy implementation?'

# 2 Wind Energy Resource Potential

How much of the world's electricity needs could actually be met using wind energy? This is a question of fundamental importance. Detractors of wind energy, and of renewable energy in general, often assert that modern renewable energy will never contribute more than a few per cent of world energy demand and is therefore not worthy of serious consideration. Is such scepticism justified?

This question can be briefly examined in two ways. First, a quick look at Denmark, which has to date pursued the world's most intensive wind energy development, reveals that in 1998 wind accounted for 9 per cent of Denmark's total electricity production. This share is set to continue growing in the future, contributing a major portion of Denmark's total electricity demand. Secondly, electricity generation is one of the world's largest industries, and global electricity demand is expected to surpass 25 000 TWh/yr (25 trillion kWh/yr)[1] around the year 2020 or 2025. If wind energy supplies only 1 per cent of this demand, then assuming a wholesale electricity price of 0.03 US$/kWh, wind energy's annual electricity production would still be worth US$7.5 billion (thousand million) per year, more than many other entire industries. Furthermore, with an installed capacity of nearly 18 500 MW in 2000 and assuming a capital cost of US$1000 per kW, the world's investment in installed wind capacity was already worth approximately US$18.5 billion in 2000. Wind energy is therefore of interest not only because it can potentially meet a large fraction of countries' electricity demand, but also because even a small fraction of the global electricity market amounts to a major industry in terms of both investment and annual revenue.

More formal analysis is necessary, however, to better understand the size and potential of the world's wind energy resource. This chapter therefore examines both the physical and practical potential of wind energy. We begin with a summary of worldwide installed wind turbine capacity to date, followed by a brief primer on wind resource assessment. These are then followed by longer-term scenarios for future wind energy utilisation, as well as the European Union's strategy for reaching its long-term wind turbine installation goals. Lastly, the chapter presents an analysis of the physical and economic feasibility of more ambitious wind power development, highlighting a Danish study to meet 50 per cent of Danish electricity demand through wind by the year 2030.

## Worldwide installed capacity

Since 1980, modern grid-connected wind turbines have been installed in more than 50 countries around the world. Early installations were predominantly in industrialised countries, with the USA and Denmark accounting for almost 90 per cent of installed capacity in the early 1990s. Though the USA dominated the field in the 1980s, its wind capacity growth rate slowed dramatically in the 1990s and was even negative for a period as old units were taken out of service and not replaced. Meanwhile, as shown in Table 2.1, Germany and Spain have experienced very dramatic wind capacity growth in the 1990s and surpassed the USA in total installed capacity. As of late 1998, however, installations have significantly picked up again in the USA, though not at the levels seen in Germany. Other major European players include the Netherlands and the UK.

A number of European countries not specified in Table 2.1 have also recently initiated significant wind energy programmes. By the end of 2000, the installed wind power capacity in these European countries included: Portugal 111 MW, Austria 69 MW, France 63 MW and Finland 39 MW (BTM Consult, 2001). Activity in developing countries has also picked up significantly in recent years, particularly in India and China. India currently ranks fifth in the world in total installed wind power capacity. Argentina, Cape Verde, Costa Rica, Egypt, Iran and Mexico are other countries not specified in Table 2.1 with notable recent wind energy growth.

**Table 2.1** Worldwide grid-connected wind capacity (MW)

| *Region* | *Country* | *1995* | *1996* | *1997* | *1998* | *1999* | *2000* |
|---|---|---|---|---|---|---|---|
| Americas | Canada | 21 | 21 | 26 | 83 | 126 | 139 |
| | United States | 1 591 | 1 596 | 1 611 | 2 141 | 2 445 | 2 610 |
| | Other American | 12 | 34 | 44 | 68 | 97 | 98 |
| Asia | China | 44 | 79 | 146 | 200 | 262 | 353 |
| | India | 576 | 820 | 940 | 992 | 1 035 | 1220 |
| | Other Asian | 10 | 13 | 22 | 33 | 79 | 157 |
| Europe | Denmark | 637 | 835 | 1 116 | 1 420 | 1 738 | 2 341 |
| | Germany | 1 132 | 1 552 | 2 081 | 2 874 | 4 442 | 6 107 |
| | Greece | 28 | 29 | 29 | 55 | 158 | 274 |
| | Ireland | 7 | 11 | 53 | 64 | 74 | 122 |
| | Italy | 33 | 71 | 103 | 197 | 277 | 424 |
| | Netherlands | 249 | 299 | 329 | 379 | 433 | 473 |
| | Spain | 133 | 249 | 512 | 880 | 1 812 | 2 836 |
| | Sweden | 69 | 103 | 122 | 176 | 220 | 265 |
| | UK | 200 | 273 | 328 | 338 | 362 | 425 |
| | Other European | 43 | 74 | 120 | 170 | 221 | 363 |
| Others | | 36 | 46 | 57 | 83 | 151 | 242 |
| **Total by end of year** | | 4 821 | 6 104 | 7 639 | 10 253 | 13 932 | 18 449 |
| **Annual growth** | | | 1 283 | 1 535 | 2 514 | 3 779 | 4 517 |

*Source*: BTM Consult (1997, 1999, 2001).

## Energy in the wind

Before discussing future wind energy potential, it is helpful to have a basic understanding of the physical properties of energy in the wind and of how to estimate wind resource availability. This section provides a brief introduction to some basic principles of energy extraction from the wind.

The kinetic energy of a volume of air $V$, moving at the speed $u$ is:

$$KE = {}^{1}/_{2}\, \rho V u^2, \qquad (2.1)$$

where:

$KE$ = kinetic energy (kg m$^2$/sec$^2$, or joules)
$\rho$ = air density (kg/m$^3$)
$V$ = volume of air (m$^3$)
$u$ = air speed (m/sec).

Power is expressed in terms of work per unit time, or in other words, the change in kinetic energy per unit time, $\frac{d(KE)}{dt}$. To obtain the expression for power, we can rewrite equation (2.1) as:

$$KE = {}^{1}/_{2}\, \rho(Area \cdot dx)u^2,$$

such that the volume of air $V$ is expressed by an *Area* perpendicular to the wind flow multiplied by the horizontal displacement in the direction of wind flow $dx$. The power, or change in kinetic energy per unit time, is then expressed by:

$$Power = \frac{d(KE)}{dt} = \frac{d}{dt}\left[\tfrac{1}{2}\rho(Area \cdot dx)u^2\right] = \tfrac{1}{2}\rho\left(Area \cdot \frac{dx}{dt}\right)u^2. \qquad (2.2)$$

Since $\frac{dx}{dt}$ is in fact the wind speed $u$, power can be expressed as

$$Power = {}^{1}/_{2}\, \rho \cdot Area \cdot u^3.$$

When seeking to extract energy from the wind, it is this power passing through the fixed area of the wind turbine rotor which is of interest. The power (or kinetic energy flux), expressed per unit of area (of the rotor), is known as the power density $P$:

$$P = {}^1/_2\, \rho u^3. \tag{2.3}$$

The power density is expressed in terms of watts per square metre. From equation (2.3) we see that the power density is a function of the cubed wind speed, meaning that an increase in wind speed by a factor of 2 leads to an increase in power density of $2^3 = 8$. This exponential relationship, between wind speed and the power which can be potentially extracted by a wind turbine, highlights the paramount importance of wind speed when selecting locations for wind power plants.

Naturally, wind speed in the atmosphere is not constant, but varies over time, expressed mathematically as $u = u(t)$. Figure 2.1 shows an example of half-hourly averages of wind speed and a wind turbine's power output over the course of 6 months.

**Figure 2.1** 30-minute averages of wind speed and wind turbine power output over 6 months

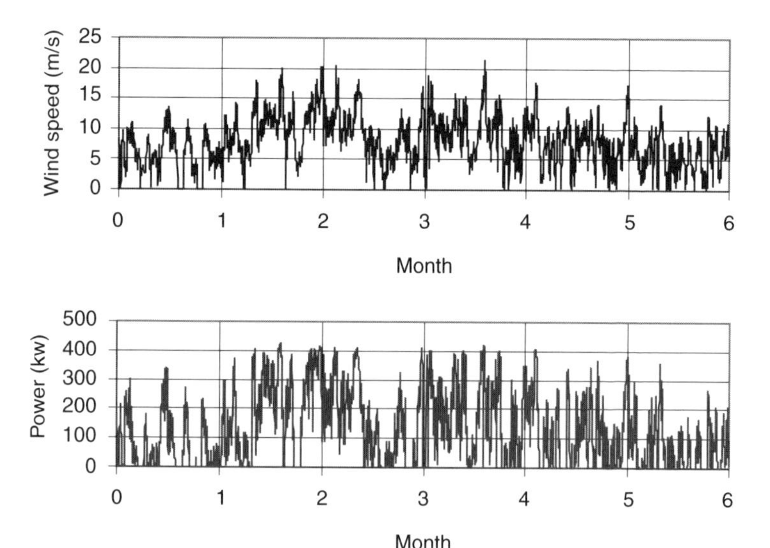

Given the variability of wind speed, a realistic measure of the available wind power resource is provided by the long-term mean power density $\bar{P}$:

$$\bar{P} = \frac{1}{T}\int_0^T \tfrac{1}{2}\rho u(t)^3 dt = \int_0^\infty \tfrac{1}{2}\rho u^3 f(u)\,du \tag{2.4}$$

where $T$ is the time over which the average is taken. $T$ should be large, such as one year, or even better, 10–20 years. This is because wind speed varies significantly during the year and even the annual average wind speed may vary by up to 10–20 per cent between different years. The function $f(u)$ is the frequency distribution of wind speed, that is, the probability of the wind speed being within a given (unit) interval at any given time.

The mathematical Weibull two-parameter frequency distribution can provide estimated wind speed probability distributions which have proven to fit well with measured wind speed data. The Weibull distribution is defined as follows:

$$f(u) = \left(\frac{k}{A}\right)\left(\frac{u}{A}\right)^{k-1} \exp\left(-\left(\frac{u}{A}\right)^{k}\right), \tag{2.5}$$

where:

$f(u)$ = the estimated frequency of occurrence of wind speed $u$;
$A$ = the scale parameter ($A > 0$);
$k$ = the shape parameter ($k > 1$);
$u$ = wind speed ($u \geq 0$).

The Weibull scale and shape parameters vary by location, depending on climate and terrain conditions. The two Weibull parameters are determined from measurements when these are available for an actual site. If no measurements are available, the Weibull parameters can be estimated through the 'wind atlas' methodology discussed subsequently in this chapter.

The Weibull shape parameter $k$ defines the shape of the wind distribution and varies with the actual climate. In typically low wind areas such as the Arctic regions and the tropics, the value of $k$ is close to 1. In climatic regions dominated by the westerlies such as in north-western Europe, the value of $k$ is approximately 2, indicating a Rayleigh distribution of wind speeds. In areas near the equator

**Figure 2.2** Measured wind speed frequency distribution (columns) and Weibull fit (line) to the measurements

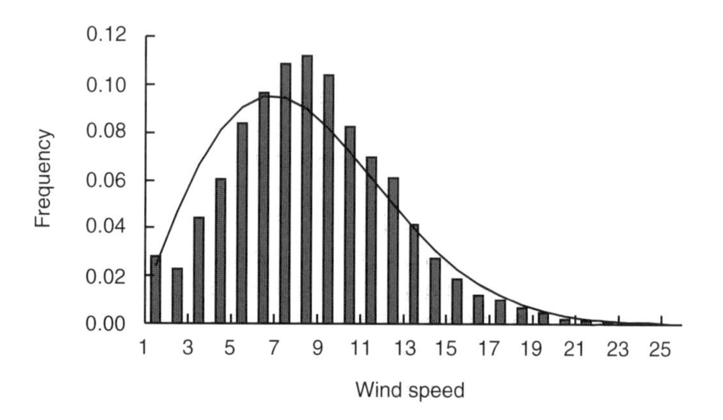


dominated by constant trade winds, $k$ can be of the order of 3 or higher, approaching a normal distribution of wind speeds.

As an approximation, the scale parameter is related to the annual mean wind speed as follows:[2]

$$\bar{u} = \frac{1}{T}\int_0^T u(t)dt = \int_0^\infty uf(u)du \approx 0.89 \cdot A. \tag{2.6}$$

A measured histogram of wind speed data is shown in Figure 2.2, together with the Weibull fit. The figure is typical of the westerlies wind regime (see following section) and represents a shape factor of approximately 2. Figure 2.2 demonstrates that, with good estimates of the scale and shape parameters, close approximations of the actual wind speed probability distribution can be obtained, allowing good estimates of mean power density.

## Wind resource assessment and data limitations[3]

The term 'wind resource assessment' is usually defined as a calculation of the average wind speed over 10 to 20 years at a specific site or area. Accurate determination of the average wind speed is of paramount importance. As discussed above, as a rule of thumb wind turbines' power output increases by the cube of the wind speed, resulting in a substantially reduced cost of generated electricity in high wind locations.

Wind resource assessment for a site or area is based on two elements: high-quality nearby wind measurements (preferably on-site) and a micro-siting model, which can estimate the spatial distribution of the wind resource over the entire area. Using only measurements from a nearby meteorological station (for example, at an airport) will cause the local effects on the air flow around that station's mast to be 'transported' to the wind turbine site in question, resulting in erroneous results. For example, if the meteorological mast is located near a building, which will reduce the wind speed of the flow coming from that direction, this reduction would almost certainly not be found at the wind turbine site. Clearly, therefore, models are necessary to obtain accurate estimates of the wind resource at any particular site.

**Wind atlas methodology**

The most widespread micro-siting models are based on the physical laws governing wind flow. An example of the physical approach is the 'wind atlas' methodology, but other models exist as well. The wind atlas methodology has been used for wind resource assessment and siting around the world and present-day state-of-the-art models are able to predict the wind resource with good accuracy in many areas.[4]

Wind speeds measured at a meteorological station are determined mainly by two factors: regional overall weather systems, which often have an extent of several hundred kilometres and the local topography around the site in question (a few tens of kilometres from the station). The wind atlas methodology (Troen and Petersen, 1989) is a comprehensive set of models for horizontal and vertical extrapolation of wind speeds measured at a meteorological station (for example, at an airport) for estimation of wind resources at a nearby site (for example, a planned wind farm).

The models are based on physical principles of flows in the atmospheric boundary layer, and they take into account: (1) terrain roughness (for example, desert surface, farmland, water surface), (2) sheltering effects (due to buildings and other obstacles), and (3) orography (terrain height variations such as hills and escarpments). Terrain roughness is often standardised into roughness classes. Roughness class 0 covers smooth surfaces such as sand or desert surfaces; roughness class 1 represents open farmland with very few buildings; roughness class 2 represents more closed farmland with some trees and/or bushes; roughness class 3 is characterised by more sheltered terrain, suburbs and so on.

For each meteorological station the wind atlas tables provide calculated Weibull *A*- and *k*-parameters for 12 sectors of the wind rose,[5] five heights and four roughness classes. In addition, the sector-wise distribution of wind speed is given in per cent for each roughness class. A summary table gives estimated annual mean wind speed and mean power density for each of the five standard heights and four roughness classes. This is illustrated in Table 2.2. Based on such information, the wind atlas methodology is able to extrapolate the wind resources from meteorological stations onto nearby wind turbine sites.

The following comprises a broad overview of wind resource assessment and siting around the world. For convenience, the world has been divided into a number of regions according to their wind climate. The characteristics of these regions are described, as is the ability of state-of-the-art models to predict these regions' wind resources.

## The Arctic

So far, the exploitation of wind power in Arctic and sub-Arctic regions has been scarce, but this may well be changing. The barriers to widespread application of wind energy in Arctic regions comprise technological, economic, social and institutional barriers. However,

**Table 2.2** Summary wind atlas table for Hurghada on the Egyptian coast of the Gulf of Suez (z refers to height above terrain, U is the estimated annual mean wind speed, and E is the estimated annual mean power density in the wind)

| z | *Roughness class 0* | | *Roughness class 1* | | *Roughness class 2* | | *Roughness class 3* | |
|---|---|---|---|---|---|---|---|---|
| (m) | U (m/s) | E (W/m$^2$) | U (m/s) | E (W/m$^2$) | U (m/s) | E (W/m$^2$) | U (m/s) | E (W/m$^2$) |
| 10 | 6.9 | 327 | 5.7 | 203 | 4.8 | 121 | 4.2 | 80 |
| 25 | 7.6 | 422 | 6.7 | 300 | 5.8 | 197 | 5.2 | 143 |
| 50 | 8.2 | 516 | 7.6 | 415 | 6.7 | 285 | 6.1 | 218 |
| 100 | 8.8 | 667 | 9.0 | 698 | 7.9 | 463 | 7.2 | 353 |
| 200 | 9.8 | 926 | 11.4 | 1 447 | 9.8 | 910 | 8.9 | 676 |

*Source*: Mortensen and Said (1996).

many of these barriers are not unique to the Arctic, or even particularly severe there. One important barrier, however, is the lack of adequate knowledge of the wind resources in candidate regions. Apart from the northern parts of Sweden and Finland, little appears to have been done with respect to a systematic mapping of Arctic and sub-Arctic wind resources. Moreover, it is not clear to what extent the methods developed for wind resource estimation and siting in the temperate climates will apply to these regions. Snow, ice and sub-zero temperatures not only make it difficult to make reliable wind measurements, but they also change the roughness of the terrain considerably from season to season. Furthermore, the cooling of the lower layers of the atmosphere leads to local wind flows to some extent. Consequently, it is often very difficult to extrapolate the measured wind climate over more than a few kilometres.

### Temperate plains and the westerlies

The temperate plains, which cover significant portions of North America, Europe and Asia, are characterised by large-scale low-pressure systems moving over these areas. These systems give rise to powerful storms and, because of the regularity of these systems, a steady wind climate. The westerlies refer to the wind regime of the northern hemisphere where wind from the west is predominant. Usually the westerlies refer to the wind regime on both sides of the North Atlantic, including, for example, eastern Canada, southern Greenland, the British Isles, the Scandinavian peninsula and parts of north-western Russia. The 'Roaring Forties' in the southern hemisphere are also a western wind belt dominated by the westerlies.

In wind energy terms, the strength and regularity of weather systems in the temperate plains and westerlies means that the energy production potential and reliability of predictions of production can be expected to be high. The wind atlas methodology was developed with these areas in mind. The method is used to estimate the expected production at a given site using wind data from up to 100 km away. Since, generally speaking, the meteorological network is very dense in these areas, the wind energy potential at virtually any location can be calculated. Furthermore, numerous studies have shown that the method gives very reliable results for most of the temperate regions, as long as the terrain is not too complex.

### Deserts and semi-arid areas

From a wind energy point of view, deserts and semi-arid regions have a number of advantages: land-use intensity is often very low, access is easy and construction work relatively simple. Also, the surface roughness of the land tends to be low and uniform, so siting can be done primarily with optimisation of power production, or minimisation of cost, in mind. Such areas could provide space for large-scale utilisation of wind energy, provided they are favoured by a healthy wind climate and located not too far from places where power is in demand. Unfortunately, as in other sparsely populated regions, the meteorological network is very coarse at present and the wind climate tends not to be known in great detail. The physics of the wind flows in these dry regions of high solar insolation and little vegetation are also quite different from the temperate regions, where most of the models and techniques for wind resource estimation and siting were developed and tested. However, studies carried out in, for example, Algeria, Libya, Egypt, Israel, Syria and Jordan should lead to a better understanding of the limits of contemporary models in these regions.

### The tropics

The tropical regions are often characterised by a high need for improved power provision, with many people still lacking access to electricity. Very high population growth is also found in these areas, creating even higher demand for electricity in the near future. As a result, there is an increasing interest in all kinds of energy, including wind.

The tropical regions are dominated by seasonal wind systems, like the monsoon and the trade winds. In many areas the measuring network is dense and dates back many years, providing long records which are highly useful for wind energy purposes. Because of the dense network, quite reliable estimates of the expected wind resource can be obtained for many tropical areas. The task is made slightly more difficult, however, by the fact that localised thermally driven wind systems can be found in some areas. Studies along the lines laid out in the European Wind Atlas have been carried out in many places. Two examples include Somalia and India, which are both dominated by monsoonal-type flows. Regional studies have verified the wind atlas methodology by comparing the predicted production of wind farms to actual production. India also has a very comprehensive database of meteorological measurements. A wind atlas has also been made on the Cape Verde islands, and again the method has been verified with good results, using actual output from wind farms.

### Open sea

Wind turbines have a significant local visual impact, and siting of wind turbines can meet with resistance in densely populated countries like the Netherlands, Denmark and the UK. As a result, siting wind turbines in the open sea or in shallow coastal waters has become increasingly attractive, as discussed in Chapters 4 and 6. The open sea is in general characterised by very high wind potential, but a detailed and reliable map of these resources is very difficult to produce due to the extremely sparse measuring network.

There are two sources of information available for estimating the offshore resource: measurements from small islands, which are few and far between, and the so-called COADS database. COADS is short for the Comprehensive Ocean–Atmosphere Data Set and is a result of continuing co-operation between several American institutions (see, for example, Diaz et al., 1992). The data set contains measurements of the wind speed and direction as reported from ships crossing the oceans. This gives a huge – albeit in some areas sparse – data set covering most of the oceans. The data set has been compared to coastal measurements in some areas and the overall agreement appears to be good. Other sources of information are available for certain limited offshore areas, including the wind atlases for the North Sea, Baltic Sea and Gulf of Suez.

### Coastal areas

Land sites near the coastline have always been in demand for wind power generation because of the generally high wind resource compared to (flat) inland sites in the same wind regime. This demand, as well as many other claims on coastal land areas, has led to a decrease in the availability of such sites; and near-coastal offshore sites have therefore become more attractive. Taking 'near-coastal offshore' to mean the offshore area where the influence of the land on wind flow is still present, this zone is on the order of 10 kilometres wide. Several conflicting constraints must be taken into account in the siting of offshore wind turbines. As discussed in Chapter 4, the cost of construction, grid connection and maintenance transport all increase with increasing distance from the coastline, but so does the available wind resource. Costs can be reduced by erecting turbines closer to the shore, but here visual impact and interference with other activities may be (too) high.

Because wind resources (and costs) vary considerably over small horizontal distances, there is an increasing demand for accurate offshore wind resource estimates. In particular, this presents a challenge to the physical models, since offshore wind measurements very rarely exist and would be very costly to obtain.

### Mountains

In mountainous regions the topography enhances the existing wind potential, resulting in very high potentials at certain sites. However, the exact magnitude of this potential is difficult to assess accurately because mountainous areas are often sparsely populated and consequently have very limited wind-measuring networks.

Because of the complex nature of the terrain – as well as the fact that the winds are often dominated by local effects, driven, for example, by local differences in temperature – it is very difficult to model wind flow in mountainous areas. A European Union initiative funded under its JOULE programme is attempting to address this problem by combining micro-siting models with models covering the wind flow over a larger area, typically hundreds of kilometres. This approach is being tested in Ireland, northern Portugal, central Italy and Crete and is indeed showing promising results in these regions.

### Wind resource estimation

This section provides a brief step-by-step approach to wind resource assessment for an area. Typically these tasks will be carried out by the national meteorological office in co-operation with wind energy consultants. As a rough estimate, the total costs of a complete study as laid out below would be around US$1 million for a country with an extensive measuring network, as can be found in most developed as well as developing countries.

#### *Overview of existing measurements*

As a first step in any resource assessment for an area, the existing measurements must be analysed. The purpose of this step is twofold.

First, it provides input for a preliminary coarse map of the wind resource for the area. Secondly, it also allows an assessment of the overall quality of the measuring network.

*Coarse map of wind resources of the area*

Once the existing measurements have been analysed, a coarse map of the area can be generated. This map is based only on the existing measurements, and it will only give a rough idea of the location of potentially high (and low) wind areas.

*A first coarse wind atlas*

Using the existing data, a preliminary wind atlas can be made. Using the wind atlas methodology, the existing measurements can be extrapolated to the whole area in question. This wind atlas will not provide a very accurate picture of the resource, but it can point towards interesting areas of high wind.

*Focus on interesting areas*

Using the coarse wind atlas, the most interesting areas can be identified for further study.

*Identifying measuring sites*

Within the interesting areas, sites for detailed wind energy-oriented measurements can be identified.

*Measure for at least 1 year, preferably 3–5 years*

Once the measuring sites have been identified, measurements should be carried out for at least one year to obtain a fair representation of the wind's annual variation, but preferably for 3–5 years to obtain some idea of the climatological variability.

*Wind atlas*

Based on the original and new measurements, a complete wind atlas can be made. This wind atlas will then form the basis of any further wind energy resource estimates. The atlas can be used not only to find interesting areas but also to provide fairly exact estimates of the actual production of a potential wind turbine site.

The wind atlas for the Gulf of Suez is an example of such a regional wind atlas (Mortensen and Said, 1996). This wind atlas was a result of a comprehensive 5-year wind resource assessment programme covering a 250-km stretch of the Gulf of Suez and the northern Red Sea. The study was based on measurements on four 25-metre masts, and in addition historical data from five existing stations were analysed. The project documented higher wind speeds than hitherto assumed, and it formed a basis for the ambitious plans for wind power development in Egypt.

### Uncertainties

Uncertainties in the prediction of a wind turbine's or wind farm's annual output depend on the quality of the data for the wind resource and for the wind turbine's power curve. For flat terrain, the standard deviation of the wind resource is typically 3–4 per cent of the average annual wind speed. This is equivalent to an approximately 5–10 per cent standard deviation on the average annual energy production. For mountainous regions, the deviation doubles. The standard deviation on wind turbine power curve is typically 5–6 per cent of annual energy production in simple terrain, 10 per cent in complex terrain and 15 per cent in very complex terrain. In north-western Europe, wind farms' annual electricity production can be predicted with an overall uncertainty of 10–15 per cent.

In some areas, however, wind energy potential can still not be satisfactorily estimated. This means that wind energy meteorology today faces two primary tasks: first, to educate users in the models currently available and in their proper use and known limitations; and secondly, to conduct research in fields where knowledge is still missing. Part of this research will involve collecting and evaluating the results of the numerous studies that have already been carried out.

## Global wind resource potential

The total solar radiation intercepted by the earth is approximately 180 000 TW-yr/yr (1.58 billion TWh/yr), corresponding to an average of 350 W/m$^2$ over the earth's surface, though this is distributed much more towards the equator and less towards the poles. In comparison, global annual electricity consumption is on the order of 1.5 TW-yr/yr. Most of the incoming radiation is lost again to outer space as outgoing radiation. A small part, on the order 3–5 per cent of incoming radiation, is converted into the kinetic energy of the moving atmosphere through the generation of global, regional and local temperature differences, forming the basis for the world's wind energy resource. Of this global kinetic energy flux, only a minute fraction can even theoretically be captured as useful wind energy, since wind energy 'extractors' can only extend a mere 100 metres or so into the atmosphere. Nevertheless, the theoretical global potential for extracting energy from the wind far exceeds the world's total energy consumption.

A number of researchers have investigated the world's technical and exploitable wind resource potential. Grubb and Meyer (1993), van Wijk et al. (1993) and the World Meteorological Organization have estimated the total global wind energy resource to be on the order of 60 TW-yr/yr (approximately 500 000 TWh/yr). Of this theoretical potential, Grubb and Meyer (1993) estimate the practical or exploitable worldwide wind potential to be approximately one-tenth, or 6 TW-yr/yr, after accounting for social, aesthetic, land use and environmental factors which will ultimately limit total wind power development.[6] The exploitable potential therefore appears to be approximately four times the current global electricity consumption. An earlier study by the International Institute of Applied Systems Analysis (Heifele et al., 1981) points to a theoretical potential of 500 TW-yr/yr and an exploitable wind resource of 3 TW-yr/yr based only on coastal regions. Therefore, even by the most conservative estimates, the total exploitable wind energy resource is approximately double the total current worldwide electricity consumption.

A more specific study in the USA by the US Energy Information Administration attempted to identify economically viable wind resources located within proximity to existing high-voltage transmission lines. This study calculated a wind energy potential of close to 1 million average megawatts within 20 miles of existing transmission lines in the USA, far greater than the country's total existing generation capacity (USEIA, 1995).

On a global basis, therefore, wind energy resource availability is not a significant issue. Wind energy's potential contribution to the world's overall electricity supply is limited not by resource availability, but by economic and social factors, as outlined in subsequent chapters. For individual countries, wind resource availability will vary based on geographical conditions. Some countries will have a large excess of wind resources, while others will be more limited. Wind resources are expected to be most abundant in the temperate zones of North America, Europe and north-central Asia. Wind resources may on average be somewhat lower in Africa, Australia and Latin America; but nevertheless, these areas also contain sizeable areas of high wind availability including, for example, much of coastal North and West Africa (see Grubb and Meyer, 1993).

## Future medium- to long-term implementation of wind power

The implementation of grid-connected wind power in the global energy system depends on several conditions, including:

- The identified physical potential for erecting wind turbines;
- the economic competitiveness of wind power compared to conventional power production;
- the need for additional electricity production capacity, including non-polluting electricity;
- barriers to be overcome, including institutional barriers and unfavourable electricity pricing structures;
- incentives for increased development and application of renewable energy.

The above discussion highlighted the abundant physical potential for wind power worldwide. Furthermore, as discussed in Chapter 4 below, the competitiveness of wind power has improved considerably since the establishment of the modern wind energy industry in the early 1980s. At present wind turbines located in high wind areas are either competitive or close to being competitive with conventional power plants in terms of total production costs. Nevertheless, many barriers to increased wind power adoption continue to exist, including financial and institutional barriers, discussed in Chapters 5 and 7.

Given these factors, what is a realistic prognosis for future wind power development? The following pages summarise a number of studies which have investigated the planned and likely future growth of wind energy in the medium- to long-term.

### European Union policy strategy

In the autumn of 1996 the EU Commission launched a Green Paper on a strategy for the development of renewable energy within the European Union (EU). The Green Paper stimulated an extensive debate on the prospects for renewable energy, and the Green Paper's publication was followed by numerous reactions from member state government agencies, industry associations, research institutes and non-governmental organisations. These reactions were incorporated into the EU Commission's subsequent White Paper *Energy for the Future: Renewable Sources of Energy* and its proposed Action Plan (European Commission, 1997).

The strategy and action plan of the White Paper present a goal of meeting 12 per cent of the European Union's gross inland energy consumption through renewable sources by the year 2010, mainly through biomass, hydro power, wind energy and solar energy. Projected energy demand is based on what is termed the 'Conventional Wisdom Scenario (European Energy to 2020)'.

At present renewable energy sources contribute less than 6 per cent of the EU's overall gross inland energy consumption, while at the same time the EU's dependence on energy imports is approximately 50 per cent and expected to rise in the coming years if no action is taken. At the December 1997 Third Conference of the Parties to the United Nations Convention on Climate Change in Kyoto, the EU committed itself to a reduction in greenhouse gas (GHG) emissions of 8 per cent by 2008–12 compared to the 1990 level. Thus there is considerable interest in increasing the use of indigenous sources of renewable energy, thereby reducing GHG emissions, increasing energy security by reducing dependence on energy imports and simultaneously contributing to job creation within the EU.

After biomass, wind energy is expected to be the main contributor of future renewable energy in the EU. The installed capacity of wind power in EU countries is proposed to grow from approximately 4.5 GW in late 1997 to 40 GW by the year 2010. If current wind energy growth rates persist, this appears to be a realistic though ambitious goal. If this 40 GW target is achieved, wind power will then cover approximately 3 per cent of total EU electricity demand in 2010, compared to less than 0.5 per cent today.

Table 2.3 summarises the main assumptions and estimates for wind energy in the White Paper. Without a determined and

**Table 2.3** Assumptions for wind energy in the EU White Paper

| *Wind energy* | *Estimates* |
|---|---|
| Additional capacity, 1997–2010 | 36 GW |
| Unit cost, 1997 | US$ 1 130 kW |
| Unit cost, 2010 | US$ 825 kW |
| Total investment, 1997–2010 | US$ 32.5 billion |
| Achieved $CO_2$ reduction, 2010 | 72 million tons/year |

*Source*: European Commission (1997). Exchange rate: 1 ECU = US$ 1.129.

co-ordinated effort to mobilise the Union's renewable energy resources, a significant portion of this potential will go unrealised. Thus, the Commission proposed an action plan to carry this development goal towards realisation. The action plan aims to provide fair market opportunities for renewable energy without imposing excessive financial burdens on society at large.

The following is a list of measures contained in the action plan, aimed at overcoming obstacles to reaching the indicated objective of 40 GW of installed wind power capacity (the list only includes those measures from the White Paper that are relevant to wind energy):

*Objectives and strategies*

- Community strategy and overall EU objective of 12 per cent renewable energy use by 2010.
- Member states set individual objectives for 2005 and 2010 and establish strategies (action).

*Internal market measures*

- Fair access for renewables to the electricity market (directive).
- Restructuring the Community framework for taxation of energy products (revised directive).
- Development and/or harmonisation concerning 'golden' or 'green' funds (action).

*Reinforcing Community policies*

- Inclusion of actions on renewables in the overall strategy for combating climate change.
- Adoption and implementation of the 5th Framework Programme for Research, Technology Development and Dissemination (1998–2002).

- Renewables to be included in the main priorities along with employment and environment in the regional fund new phase (2000–2006).
- Examination of adequacy of existing instruments and possibility of further harmonisation (Agenda 2000 review).
- Definition of an energy strategy for co-operation with African, Caribbean and Pacific States in the Lome Convention Framework, emphasising the role of renewables.

*Strengthening co-operation between member states*

*Support measures*
- Development of European standards and certifications.
- Better positioning for renewables among institutional lenders and the commercial market by developing schemes for facilitating investment in renewable energy projects.
- Creation of a virtual centre 'AGORES' for collection and dissemination of information.

*Campaign for take-off*
- 10 000 MW of large wind farms (co-funding).
- Integration of renewable energy in 100 communities (co-funding).

*Follow up*
- Scheme to monitor progress.
- Improvement of data collection and statistics.
- Inter-services co-ordination group.
- Creation of working group involving Commission and member states.
- Regular reporting to the Union's institutions.

One of the key actions in the campaign for take-off is the proposal for 10 000 MW of large wind farms, which will represent approximately 25 per cent of the feasible overall wind energy development goal for 2010 outlined in the White Paper. Establishment of these wind farms will receive co-funding from the EU. The remaining 30 000 MW of the overall target are not expected to require public funding provided that fair access for wind turbines to the European grid is guaranteed.

**Table 2.4** Development of the world market for wind turbines

| *Year* | *Annual installed capacity (MW)* | *Growth rate of annual installed capacity (%)* | *accumulated capacity (MW)* | *Growth rate of accumulated capacity (%)* |
|---|---|---|---|---|
| 1992 | 231 | – | 2 278 | – |
| 1993 | 480 | 108 | 2 758 | 21 |
| 1994 | 730 | 52 | 3 488 | 27 |
| 1995 | 1 290 | 77 | 4 778 | 37 |
| 1996 | 1 292 | 0.2 | 6 070 | 27 |
| 1997 | 1 566 | 21 | 7 636 | 26 |
| Average growth | | 47 | | 27 |

*Source:* BTM Consult (1998a).

### Short- to medium-term development

BTM Consult, a Danish wind energy consultancy, has recently evaluated the prognosis for short- to medium-term development of world-wide grid-connected wind power capacity (BTM Consult, 1998a).

Since 1992, the wind turbine market has developed substantially. The annual sale of wind turbines has increased significantly, and the accumulated global wind power capacity has increased by 27 per cent per annum in the period 1992–7,[7] as shown in Table 2.4.

The BTM Consult study concludes that the existing global installed capacity of 7.6 GW in 1997 is projected to grow to approximately 20 GW by the year 2002, a growth rate of more than 20 per cent per year. This projection is based on recent experiences and trends in the most important wind turbine markets. Development in India, China, USA, Germany, Spain and Denmark is expected to play a particularly important role. In the short term, no capacity constraints in the manufacturing industry are foreseen. On the contrary, strong competition between manufacturers is envisaged. Looking ahead to the year 2007, total global accumulated wind turbine capacity is estimated to increase to 46 GW, amounting to a growth rate of 18 per cent per year between 2002 and 2007.

### World Energy Council: long-term development

In the early to mid-1990s, the World Energy Council (WEC) prepared two global scenarios on the penetration of new renewable energy resources, looking ahead to the year 2020, and specifically addressing the development of wind energy (WEC, 1994). The scenarios take as starting points two global scenarios formerly developed by the WEC Commission.

*The 'current policies' scenario*

In this scenario the existing general economic and technological trends are assumed to continue. The scenario is mainly based on 'Case B' of the WEC Commission, including moderate levels of economic growth and technological development, and significantly increased reliance on energy conservation compared to a standard business-as-usual approach.

*The 'ecologically driven' scenario*

In this scenario economic growth follows the level of the Current Policies scenario, but a strong political effort towards international equity and environmental protection is assumed. The use of policy measures such as environmental taxes and $CO_2$ constraints imply considerable improvements in energy intensities and reduced $CO_2$ emissions. The scenario is mainly based upon 'Case C' of the WEC Commission.

Table 2.5 highlights the WEC's projected global energy and electricity demand in 2020 according to the two scenarios. The table highlights the considerably slower development of energy and electricity demand foreseen in the Ecologically Driven scenario compared with the Current Policies scenario.

In each of these scenarios the development of wind power was examined specifically. The analysis was carried out at the

**Table 2.5** Development of energy and electricity demand by 2020 according to the Current Policies and Ecologically Driven WEC scenarios

| *Scenario* | *Global energy demand in 2020 (EJ)* | *Global electricity demand in 2020(TWh)* |
|---|---|---|
| Current policies | 570 | 25 625 |
| Ecologically driven | 485 | 20 275 |

*Source*: WEC (1994).

regional/country level to estimate the possible penetration of wind power. The methodological approach was as follows.

1. A global investigation of wind resources and a global inventory of the electricity production system are performed for each country or region.
2. A cost comparison of wind power generation cost vs. the cost of electricity production from conventional sources is carried out. It is assumed that a substantial penetration of wind power will take place when the long-run marginal cost (LRMC) of wind power is lower than the LRMC for conventional electricity production plants.
3. If wind power achieves substantial penetration, it is assumed that this adoption will nevertheless be limited by other constraints including financial barriers and/or constrained growth rates in wind turbine manufacturing and installation capacity.

The development of wind power's generation cost thus constitutes an important assumption in both scenarios. In the Current Policies scenario it was assumed that R&D activities would not be intensified. Therefore, the assumed reduction in wind power generation costs is a very modest 15 per cent during the period 1990–2010. In the Ecologically Driven scenario it was assumed that R&D activities are intensified, and as a result the wind power production cost was forecast to decrease by 30 per cent in the period 1990–2010, and by a further 10 per cent in the period 2010–20. Both assumptions appear highly modest compared with actual developments to date. Between the early 1990s and late 1990s, per-kWh costs of wind energy have in fact already dropped by over 30 per cent; and since the late 1980s the decrease has been almost 45 per cent (see Chapter 4).

Compared to the Current Policies scenario, the Ecologically Driven scenario assumes faster development of wind turbine efficiency, imposition of a substantial carbon tax on fossil fuels, and less severe financial constraints on the development of wind energy. The overall results of the two scenarios for wind energy are shown in Table 2.6, where it can be seen that the two scenarios diverge substantially. The volume of wind-generated electricity is 2.5 times higher in the Ecologically Driven scenario than in the Current Policies scenario in 2020. In 2020 wind-generated electricity is pro-

**Table 2.6** Installed wind turbine capacity and wind-generated electricity in WEC scenarios

| | *Projected global wind capacity in 2005* | *Projected global wind capacity and electricity production in 2020* | | *Share of global electricity demand in 2020* |
|---|---|---|---|---|
| *Scenario* | *GW* | *GW* | *TWh* | *%* |
| Current policies | 62 | 180 | 376 | 1.5 |
| Ecologically driven | 83 | 474 | 967 | 4.8 |

*Source:* WEC (1994).

jected to meet approximately 4.8 per cent of total global electricity demand in the Ecologically Driven scenario, compared to approximately 1.5 per cent in the Current Policies scenario. A relatively larger share of global electricity demand is met through wind in the Ecologically Driven scenario due to a lower projection of overall electricity consumption in this scenario (through improved energy consumption efficiency).

### Long-term high wind growth scenario

In the autumn of 1998, BTM Consult carried out a study on the long-term global development of wind energy (BTM Consult, 1998b). The scope of the study was to assess whether a target of 10 per cent of annual global electricity demand could be supplied by wind power, and, if so, how soon this could realistically be achieved. BTM Consult employed a scenario approach, based primarily on the following variables: growth of global electricity demand, availability of wind resources worldwide, an evaluation of current wind technology and its manufacturing industry, and prospective future improvements.

Assumptions of the development of electricity demand were based on forecasts performed by the International Energy Agency (IEA), using the 'Energy Saving' case in its 1996 World Energy Outlook. The more efficient use of energy projected in this case is likely to be present in any strategy emphasising intensive wind energy development. However, the IEA case only covers the time period up to 2010. BTM Consult therefore extended the case until 2025 by

**Table 2.7** Projected global electricity demand – extended IEA forecast

| *Year* | *Global electricity demand (TWh)* |
|---|---|
| 2010 | 18 230 |
| 2025 | 25 264 |

*Source*: BTM Consult (1998b).

assuming a constant growth rate in electricity demand equal to the average projected growth rate between 1992 and 2010 (2.2 per cent per year). The resulting projected global electricity demand, shown in Table 2.7, is roughly compatible with the projected demand in the WEC scenarios discussed previously.

Based on this assumed development in global electricity demand, two scenarios were analysed.

*The 'recent trends' scenario*

This scenario is an optimistic business-as-usual case, where the existing positive trends in wind energy development are continued. It assumes that the experiences of those countries who have achieved the most significant adoption of wind energy (Germany, Denmark and India) are spread to other countries over time. Liberalisation of electricity markets is assumed to take place, and fair access to these markets is secured for wind energy. Fixed payment agreements between wind turbine owners and utilities prevail, and funds for improved technology transfer to developing countries are increased.

*The 'international agreements' scenario*

This scenario contains the same general assumptions as the Recent Trends scenario, but firm international commitments are assumed to further promote the adoption of wind energy. Concerning greenhouse gas (GHG) emissions, binding international agreements with fixed and quantified targets are assumed for all countries/regions signing the Kyoto Protocol of the United Nations Framework Convention on Climate Change (UNFCCC). Additional funds for R&D and technology transfer to the developing world are assumed to be made available. Models of emissions trading and joint implementation are assumed adopted.

**Figure 2.3** Wind power adoption based on two high-growth scenarios

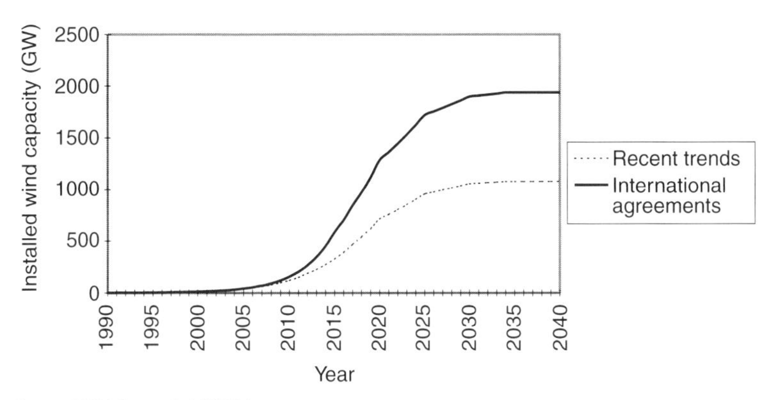


*Source*: BTM Consult (1998b).

Both scenarios assume a cumulative installed wind power capacity of 20 GW by 2002 as the starting point. The two scenarios, together with analyses of wind technology's progress ratio along the technology 'learning curve', as well as saturation levels for wind power installation, lead to calculated growth rates for cumulative installed wind power capacity. The scenario analyses were not based on modelling tools, but were carried out using fairly simple spreadsheet calculations.

Figure 2.3 shows the resulting global adoption of wind power in the two scenarios. To reach the objective of meeting 10 per cent of global electricity demand, wind power will have to supply approximately 2000 TWh per annum within 15–20 years, corresponding to approximately 900 GW of installed wind power capacity. This level of penetration could be achieved in the International Agreements scenario by 2016–17 and in the Recent Trends scenario by 2025–26.

Both scenarios indicate reductions in the production cost of wind-generated electricity from today's level to approximately 3 US cents/kWh over the next 20–25 years. This would probably make wind power fully economically competitive with conventional power production.

The possible contribution of such wind power development to carbon dioxide emission reductions is shown in Table 2.8. The Kyoto Protocol of the UNFCCC is expected to result in a global reduction in $CO_2$ emissions of 5.2 per cent (0.775 Gton $CO_2$) by

**Table 2.8** Possible contribution of wind power to reduction of $CO_2$ emissions

| *Scenario* | *Global $CO_2$ emission reductions through wind energy (Gton $CO_2$/year)* | | *Contribution toward Kyoto targets in 2010* |
|---|---|---|---|
| | *Year 2010* | *Year 2025* | *%* |
| Recent trends | 0.178 | 1.407 | 23 |
| International agreements | 0.232 | 2.529 | 30 |

*Source*: BTM Consult (1998b).

2010. The Recent Trends scenario would result in wind energy contributing approximately 23 per cent of this Kyoto Protocol target by the year 2010, while in the International Agreements scenario wind energy would contribute approximately 30 per cent. Over the long term, wind power could potentially become one of the most important options in combating global climate change.

## Large-scale implementation of wind power

Being an inherently variable resource, whose output at any given time is difficult to predict, wind power entails certain added complexities regarding integration into the electricity grid, compared to conventional electricity generation technologies. These complexities are discussed from a technical perspective in Chapter 3 and from a financial perspective in Chapter 5. Utility planners often assume that wind energy could not provide much more than 10 per cent of total electricity requirements without impacting on the technical stability of the electricity grid. Denmark, however, has been investigating the possibility of meeting a considerably higher proportion of its electricity needs through wind. This section summarises the results of one such investigation which confirms the feasibility of more intensive reliance on wind energy.

The study examined options for large-scale utilisation of renewable energy for power and heat production in the future Danish energy system and was carried out as a collaboration between Risø National Laboratory and the Danish electric utilities ELKRAFT and ELSAM. The study addressed technical and system development challenges which would arise if regional renewable energy resources are to form the main energy inputs to the future Danish power and heat supply system by 2030. Based mainly on fluctuating inputs from renewable energy technologies such as wind, photovoltaics and biomass, supply strategies were investigated which would be capable of providing the same quality of electric service as exists today (Nielsen, 1994).

Wind power's fluctuating nature means that low wind speeds and hence low power generation may occur at times of peak electricity demand. Conversely, there could be an over-supply of wind-generated electricity at times when demand is low. This is illustrated in Figure 2.4, which shows a hypothetical situation where wind power production corresponds to 50 per cent of the yearly electricity demand in Denmark. The upper curve shows the varying electricity demand in one-hour time steps during a two-week period in springtime. The fluctuating wind power production, shown as the lower curve, is based on power curves for the average wind capacity in an assumed future system. The assumed wind speeds are based on synchronous measurements at four selected locations in Denmark. In order to meet 50 per cent of the country's annual electricity demand with wind (and given wind power's low capacity factor[8]),

**Figure 2.4** Hourly electricity demand and wind power production: wind power penetration = 50 per cent

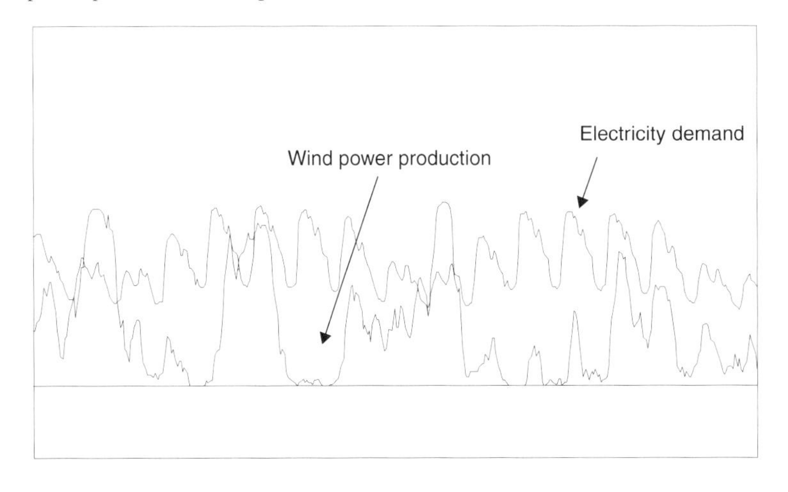

the assumed installed wind power capacity is close to the peak power demand in the system.

Figure 2.4 not only shows very substantial fluctuations in wind power production, but also that wind power production greatly exceeds total electricity demand in certain periods. This excess electricity production is further exacerbated by constraints in other parts of the power production system which require that certain other generators continue to operate during this time. A high dependence on fluctuating wind energy therefore imposes strict requirements on the regulation capability of the rest of the electricity supply system.

Several possibilities exist for addressing the problem of excess electricity production. Regulating down the wind power production during periods of high wind speeds is one possibility, while exporting electricity to other countries may be another option. Other possibilities include more flexible operation of the country's combined heat and power system. Denmark possesses a large district heating network, and much of the country's space and water heating needs are met through combined heat and power (CHP) plants which both generate electricity and provide heat for the district heating system. Options for absorbing excess electricity through the district heating system are described later in this section.

### Approach and assumptions

The study employed a scenario approach, where the basic aims of society at large formed the starting point for the analysis. Economic growth, fuel price developments, energy demand and energy supply strategies were derived in accordance with meeting these fundamental aims for the overall society.

A main scenario, called 'The Green Society', was developed which formed the basis for fundamental assumptions regarding the large-scale utilisation of renewable energy. This scenario implies, for the energy sector, an assumption of a persistent political willingness to promote energy conservation and use of renewable energy resources, with an essential goal being to achieve substantial $CO_2$ emission reductions. The analysis examined both medium- and long-term perspectives, focusing on the years 2005 and 2030. A primary goal of the 'The Green Society' is to achieve a renewable energy utilisation covering 75 per cent or more of the expected

**Table 2.9** Supply strategies for utilising renewable energy resources in year 2030

| *% of total* | *Supply strategy* | | |
|---|---|---|---|
| *Danish electricity demand* | *S1* | *S2* | *S3* |
| Wind Power | 50% | 25% | 50% |
| Photovoltaics | 0% | 0% | 15% |
| Biomass | 25% | 50% | 35% |
| **Total** | 75% | 75% | 100% |

*Source*: Nielsen (1994).

Danish electricity demand in 2030. A milestone towards this goal is to reach a 25 per cent coverage of the electricity demand in 2005 from renewable energy sources equally divided between wind power and biomass.

Three long-term electricity supply strategies for utilising renewable energy were developed, shown in Table 2.9. The S1 and S2 strategies place the primary emphasis on wind power and biomass, respectively, and both strategies aim to cover 75 per cent of total electricity demand through renewables in 2030. The third strategy, S3, includes photovoltaics as well as wind and biomass and aims to cover the entire Danish electricity demand in 2030 using renewables.

A number of models were used to carry out the analyses. These include a scenario model for energy, economic and environmental analysis of the overall system, supply system simulation and optimisation models, and a model for dynamic load flow analysis of the electricity grid.

The study carried out an assessment of the development of wind technology. Improved design and efficiency were assumed to reduce the specific costs of electricity from wind turbines by approximately 25 per cent from 1994 levels by 2030. The unit size of typical mass-produced wind turbines was assumed to increase from the 0.5 MW level available at the time of the study to approximately 2.5 MW in 2030. Furthermore, future wind turbines were assumed to operate at maximum efficiency over a wide wind speed range using variable speed and active pitch control. The installation of wind power capacity during the period up to 2030 was assumed to follow a steady path.

It was also assumed for the technical analysis that interaction with the electricity systems of neighbouring countries would be kept at current levels. Thus, the need for increased regulation capability in the system, due to large quantities of fluctuating wind power, was assumed to be met from within the Danish system, rather than relying on extensive electricity imports and exports.

### Wind power and excess power generation

Excess electricity production (electricity generation in excess of total electricity demand) increases as reliance on fluctuating wind power increases. Excess generation may further increase due to system constraints and limited regulation capability in other parts of the overall electricity production system. Figure 2.5 demonstrates excess electricity production as a consequence of increasing wind power production under Danish conditions.

When wind power generation exceeds approximately 20 per cent of total annual electricity demand, excess electricity production begins to emerge. This is demonstrated by the lower curve in Figure 2.5. When wind-based electricity covers approximately 50 per cent of total annual electricity demand, excess production will be close to 10 per cent, increasing to approximately 40 per cent when

**Figure 2.5** Excess electricity production and residual demand for conventional electricity, corresponding to different levels of wind energy penetration

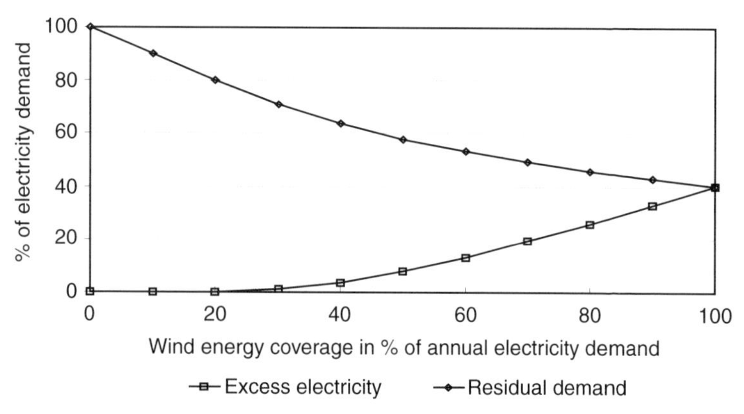


*Source*: Nielsen (1994).

electricity production from wind power equals total annual electricity demand. The upper curve of Figure 2.5 shows the percentage of electricity demand that is not met by wind power and which thus must be produced by conventional power plants. In other words, even if wind power annually produces electricity corresponding to 100 per cent of annual electricity demand, the timing of wind power generation will not fully correspond with the timing of demand. Hence, if annual wind power output equals 100 per cent of annual electricity demand, only approximately 60 per cent of this wind-generated electricity could be directly utilised, requiring that approximately 40 per cent of electricity demand be supplied by other means.

**Power supply and regulation capability**

The above discussion highlights the need for flexibility in the non-wind generation system to absorb the fluctuations of wind power output. The desired combination of high regulation capability and high efficiency of electricity production points in favour of gas-fuelled technologies. Gas technology was therefore assumed to play an important role in the future energy system, where high energy efficiency and system flexibility are essential.

In 'The Green Society' scenario, the main new technologies assumed to be introduced in the longer term are biomass gasification, integrated gasification combined cycle (IGCC) and fuel cells using natural gas and syngas.[9] Based on these technologies, the biomass utilisation in the system is expected to yield high efficiencies in electricity generation. In the short- to medium-term transition period, combined cycle plants using natural gas and circulating fluidised bed (CFB) boiler plants or multi-fuel plants utilising biomass are assumed. Gas turbines are assumed to supply peak load generating capacity. The consumption patterns for natural gas impose strong requirements on the flexibility of the gas supply system.

Heat storage (of approximately one-day capacity) is utilised to decrease or eliminate constraints on combined heat and power production. Furthermore, the heat storage capacity is used in combination with heat pumps. Excess electricity production from wind turbines is partly recovered by heat pumps to supply the district heating systems. If further heat production is required, the heat

**Figure 2.6** The percentage of different technologies covering electricity demand and 'excess' electricity utilisation

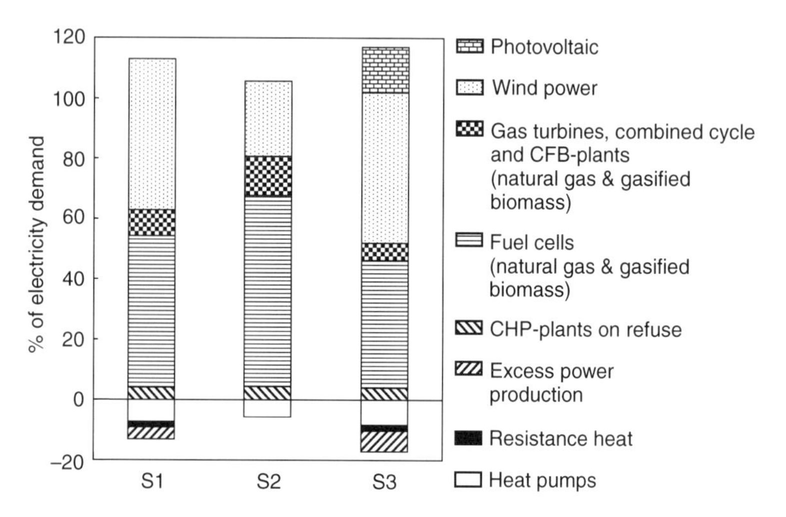


*Source*: Nielsen (1994).

pump capacity is used and electricity production is raised to supply the heat demand. The interlinked operation of the Danish electricity and heating systems thus provides a significant degree of flexibility.

Figure 2.6 shows how different technologies contribute to electricity supply in the three supply strategies, S1–S3. Wind power plays a key role in each strategy, providing 25–50 per cent of total electricity demand, as outlined earlier in Table 2.9. Excess electricity and additional electricity production to operate heat pumps are included both above and below the x-axis. Approximately half of the excess production in the S1 wind strategy and in the S3 strategy (wind, biomass and photovoltaics) is consumed by the heat pumps and in the S2 biomass strategy all of the excess production is absorbed by heat pumps.

What remains of the excess electricity production is highly irregular in time and fluctuates greatly in power. A fraction of this is recovered as resistance heat and the remainder is unusable and is lost. Such losses occur in strategies S1 and S3. In practice, this unusable excess production would not in fact be generated. The wind turbine capacity in such situations must be capable of regulating down its production to maintain stability on the national grid. However, as indicated in Figure 2.6, the percentage of total electricity production lost through excess production is fairly small, amounting to approximately 6 per cent in S1 and 8 per cent in S3.

**Results**

The achievable $CO_2$ emission reductions in the energy system as a whole and in particular in the combined heat and power sector are substantial. In the year 2030, $CO_2$ emissions from the power/CHP sector are reduced in strategies S1 and S2 by approximately 85 per cent and 88 per cent, respectively and by 100 per cent in S3 relative to the 1992 level, assuming that the burned refuse is $CO_2$–neutral. For the energy system as a whole, emissions in 2030 are reduced by 60–70 per cent from the 1992 level for all strategies, which also reflects the effects of energy conservation measures in 'The Green Society'.

The main conclusion of the technical analysis is that it should indeed be possible to develop well-functioning power and heat supply systems by 2030, in which 75–100 per cent of the electricity supply is based on Danish renewable energy resources. However, the average production cost of electricity in 2030 is expected to increase by around 30 per cent in strategies S1 and S2 and by around 65 per cent in S3, relative to the 1992 level. The composition of average production costs in the three strategies shifts towards increased investment costs and reduced fuel costs.

This conclusion regarding future electricity costs under the three strategies is based on a number of assumptions, and it must be emphasised that considerable uncertainty is associated with such long-term analyses, for example, concerning available energy resources, technological development and economic growth. It should also be mentioned that investments to improve energy efficiency on the demand side (power and heat consumption) are not included in the calculated costs on the supply side; and supply technology development costs which are not reflected in the assumed investment costs are not otherwise included in the calculated average electricity production costs. Nevertheless, the study provides encouraging signs that a future electricity scenario based on wind energy meeting 50 per cent of society's total electricity needs is neither unachievable from a system operation perspective nor altogether unreasonable from a cost perspective.

# 3
# Wind Turbine Technology and Industry

This chapter provides an introduction to wind turbine technology, a discussion of technological development and grid interaction issues, and an overview of the wind turbine industry. We begin with a brief introduction to the history of wind power use, followed by an introduction to the physical principles of extracting energy from the wind.

## A brief history of wind power utilisation

People have used technology to transform the power of the wind into useful mechanical energy since antiquity. Along with the use of water power through water wheels, wind energy represents one of the world's oldest forms of mechanised energy. Though solid historical evidence of wind power use does not extend much beyond the last thousand years, anecdotal evidence suggests that the harnessing of mechanised wind energy pre-dates the Christian era.

The use of wind power is said to have its origin in the Asian civilisations of China, Tibet, India, Afghanistan and Persia. The first written evidence of the use of wind turbines is from Hero of Alexandria, who in the third or second century BC described a simple horizontal-axis wind turbine. It was described as powering an organ, but it has been debated as to whether it was of any practical use other than as a kind of toy. More solid evidence indicates that the Persians were harnessing wind power using a vertical-axis machine in the seventh century AD (Shephard, 1990).

From Asia the use of wind power spread to Europe. Historical accounts date the use of windmills in England to the eleventh or twelth century. Witnesses also spoke of the German crusaders bringing their windmill-making skills to Syria around AD 1190. From this, one can assume that windmill technologies were generally known around Europe from the Middle Ages on. Early windmills and water wheels were used for simple low-energy processes such as water pumping and grain grinding; and they continue today to be used for this purpose in many parts of the world, particularly in developing countries. Variations in windmill styles developed from place to place, with perhaps the most famous being the traditional Dutch style. Several Mediterranean islands are also known for their picturesque old windmills.

With the advent of the steam engine in the eighteenth century the world's demand for power gradually shifted to techniques and machines based on thermodynamic processes. The advantages of these machines over wind became particularly evident with the introduction of fossil fuels such as coal, oil and gas. The advantages of steam engines and steam and gas turbines were threefold. First, the new machines were more compact and able to deliver power on a much larger scale than necessary for just water pumping and grinding, allowing a whole new level of industrial development. Secondly, the engines and turbines could be located virtually anywhere, unlike windmills and water wheels which were dependent on the availability of good sites. And third, the new machines provided more reliable power than the wind, whose availability was vulnerable to changing weather conditions.

As a result, the importance of wind energy declined during the nineteeth century and even more so during the twentieth century. The new fossil fuel-driven machines also had their drawbacks, however, because they required an external fuel source, and concentrated large amounts of power in one centralised location, making them less suitable for remote low-density locations. As a result, wind energy was able to maintain its viability in certain markets. In countries with populations scattered over large areas such as the Americas, Australia and Russia/USSR, wind power continued to play a role, particularly in the farming sector.

The traditional windrose model – the multi-bladed wind turbine used on farms throughout the world – was further developed and refined over the years. The wood used in most parts of these machines was replaced by iron and steel. Lattice steel towers were introduced, and even steel blades came into use. This transformation from wood to steel did not happen overnight but rather went on for some decades and contributed to the optimisation of these wind turbines. By the middle of the twentieth century, the Aermotor Company of Chicago claimed to have 800 000 windmills in service, mostly for water pumping. These machines were built from the late 1890s and were made of steel.

With the increasing electrification of the industrialised world, the role of wind power continued to decrease further. Fossil fuels demonstrated their competitive advantage in providing electrical power cost-effectively on a large scale. However, work on wind turbines continued to a wider extent than is commonly assumed. Though it is often assumed today that interest and research in wind power vanished due to overwhelming competition from fossil energy sources, this is in fact not the case. Around the world, theorists and practitioners continued to design and construct electricity-producing wind turbines throughout the twentieth century.

In 1891 Poul la Cour and a team of scientists at Askov Folk High School in Denmark installed the world's first electricity-producing wind turbines and established a test station for wind turbines, funded by the Danish government. As a result of this and the fuel shortage during the First World War, by 1918 one-quarter (120) of all Danish rural power stations used wind turbines for generating electricity. These turbines had a rated capacity of 20–35 kW. Also, during the Second World War, 50–70 kW wind turbines were installed in Denmark. In America the Jacobs brothers manufactured battery-charging wind turbines in the 2.5–3 kW range in large numbers from 1925 to 1957. The famous 1250 kW Smith–Putnam wind turbine was erected in 1941 at a place called Grandpa's Knob in Vermont, USA. Also, in the 1920s and 1930s the Frenchman F. M. Darrieus and the Finn S. J. Savonius designed and tested new concepts for vertical-axis wind turbines (VAWT). VAWT designs have never succeeded in gaining significant market share, but the North American firm Flowind did mass-produce a turbine of the Darrieus concept during the 1980s.

On the theoretical research side as well, efforts have continued throughout the twentieth century. La Cour carried out groundbreaking empirical observations using a primitive wind tunnel around the turn of the century. One of la Cour's students was J. Juul, who was employed by the power utility SEAS and after the Second World War headed a research and development programme on wind energy utilisation. This R&D effort formed the basis for Juul's pioneering design of the modern electricity-producing wind turbine – the 200 kW Gedser turbine – installed in 1957 and in operation until 1967.

In the 1920s the German professor Albert Betz, of the German aerodynamics research centre in Göttingen, made groundbreaking theoretical studies on wind turbines in the light of modern research. Also in the 1920s H. Glauert provided an aerodynamic theory for wind turbines. These theoretical contributions of Betz and Glauert remain the foundation of today's rotor theory, as discussed in the following section.

Other important contributors to the development of wind power theory include the Austrian engineer Ulrich Hütter, who worked in the late 1930s as chief engineer at the state-owned Ventimotor wind turbine firm in Weimar, outside Berlin. In 1942 he received his doctoral degree from the University of Vienna through a theoretical study on wind turbines; and in the 1970s he was called upon again by the West German government to lead a research effort in wind power techniques.

Important American wind energy pioneers include Palmer C. Putnam, the man behind the wind turbine at Grandpa's Knob. In 1948 Putnam issued a textbook on wind energy which is now considered a classic. Percy H. Thomas, Putnam's colleague on the Grandpa's Knob project, was also very active in this field during the 1940s. In 1955 another American, E. W. Golding, issued a textbook with the title *The Generation of Electricity by Wind Power*, and it was widely used in new editions during the 1970s and 1980s. Research and production of electricity-producing wind turbines continued in the USSR as well. In the 1950s E. M. Fateyev published a number of titles, of which at least one was translated into English by the National Aeronautics and Space Administration (NASA) and has been widely referred to.

Major international conferences included a United Nations Educational, Scientific and Cultural Organization (UNESCO) conference on wind and solar energy in New Delhi in October 1954, and a World Power Conference held in Brazil in July 1954 (Golding, 1976). In 1961 the United Nations Conference on New Sources of

Energy was held in Rome. The proceedings from this conference were published in 1964 (United Nations, 1964), and they contain key sources of information regarding the international development of wind power utilisation in the first half of the twentieth century.

Hence, research in wind power utilisation did not die because of competition from fossil fuels, but rather made steady progress over the past 100 years. The revival of more widespread interest in wind power after the oil crises of the 1970s did not require starting from scratch and was able to build on a solid foundation of theories and practical experiences. By the time the new era of wind energy began in the 1970s and 1980s, new materials and technologies had also become available. As composite materials such as fibreglass proved highly suitable for wind turbine rotor blades, blade design has become increasingly sophisticated; and electronic controls for wind turbines also continue to advance.

## Extracting energy from the wind

A basic understanding of the theoretical possibilities and limitations for extracting energy from the wind is helpful for understanding the fundamentals of wind power technology. The deductions of Betz (1920), though not directly applicable to practical engineering computations today, help illustrate the forces at work around a wind turbine propeller and are highlighted here in a slightly altered form. Figure 3.1 shows a wind turbine with a rotor radius (blade length) $r_R$, exposed to a uniform, non-turbulent flow. The undisturbed velocity has a magnitude $u_0$ and a direction perpendicular to the rotor. Behind the rotor, a circular wake with a uniform speed deficit $au_0$ expands. In other words, $a$ represents the fractional loss of wind speed through the rotor. By assumption, the wake at the point of creation has a radius equal to the rotor radius $r_R$, increasing to $r$ some distance downstream. Outside of this area impacted by the wind turbine, the wind speed is assumed to have the free stream value $u_0$.

A cylindrical control volume is devised so that it starts in the undisturbed upstream flow (to the left in Figure 3.1) and has a radius, $r$, coinciding with the wake radius, where it ends (to the right). Thus, the flow speed is $u_a = u_0 - au_0 = u_0\ (1 - a)$ at the right

**Figure 3.1** Control volume for momentum and energy balance

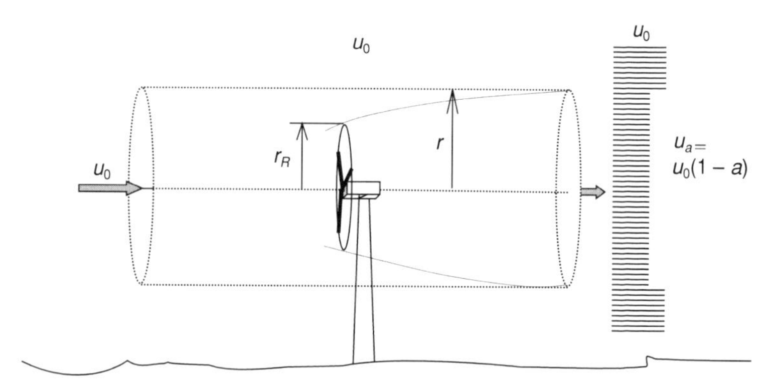


end of the cylinder, and the uniform flow speed outside the control volume is $u_0$.

As described in Chapter 2, the kinetic energy of the wind is expressed by:

$$KE = {}^1/{}_2\, \rho V u^2, \tag{3.1}$$

where:

$KE$ = kinetic energy (kg m$^2$/sec$^2$, or joules);
$\rho$ = air density (kg/m$^3$);
$V$ = volume of air (m$^3$);
$u$ = air speed (m/sec).

The volume of air $V$ in the cylinder is equal to the cross-sectional area $\pi r^2$ multiplied by the horizontal displacement, $dx$. By setting the horizontal depth of the air volume equal to the distance travelled by $u_a$ per unit of time $dt$, the horizontal displacement $dx$ equals $u_a$. Thus, the volume of air $V$ is equal to $\pi r^2 u_a$, and the kinetic energy of the air volume at any given time is equal to ${}^1/{}_2\rho\pi r^2 u_a u^2$.

As outlined in Chapter 2, power equals the change in kinetic energy over time. At the left end of the cylinder in Figure 3.1, the wind speed is $u_0$, and the kinetic energy is ${}^1/{}_2\rho\pi r^2 u_a u_0^2$, while at the right end of the cylinder, the wind speed is $u_a$, and the kinetic energy is ${}^1/{}_2\rho\pi r^2 u_a u_a^2$.

Therefore, the power extracted by the wind turbine, represented by the change in kinetic energy through the cylinder, is given by:

$$P_{wt} = {}^1/_2 \rho \pi r^2 u_a (u_0^2 - u_a^2). \qquad (3.2)$$

By convention, power is often expressed in terms of the free wind speed $u_0$, the swept rotor area, defined as the area of the circular disc 'drawn' by the blade tips ($\pi r_R^2$), and the so-called power coefficient $c_P$, representing the fraction of the wind's kinetic energy extracted by the turbine. Expressed in this way, power is given by:

$$P_{wt} = {}^1/_2 \rho \pi r_R^2 c_P u_0^3. \qquad (3.3)$$

In order to finalise Betz's deductions, we make the additional commonly held assumption that the speed deficit of the flow when passing through the rotor is half of what it finally becomes downstream. With this assumption, continuity in the wake stream tube[1] (starting with the rotor disc to the left and coinciding with the end of the control volume to the right, see Figure 3.1) yields

$$\pi r_R^2 u_0 (1 - \tfrac{1}{2} a) = \pi r^2 u_0 (1 - a) \Rightarrow \frac{r_R^2}{r^2} = \frac{1 - a}{1 - \tfrac{1}{2} a}. \qquad (3.4)$$

Combining equations (3.2)–(3.4) allows the derivation of the following expression for the power coefficient $c_P$ defined in equation (3.3):

$$c_P = {}^1/_2 (2 - a)^2 a. \qquad (3.5)$$

Again, $a$ represents the fractional loss of wind speed through the turbine. Since $a$ is unknown, this result does not appear very useful in determining the potential power yield from a wind turbine. However, by differentiating equation (3.5) with respect to $a$, the upper limit of the power coefficient can be determined:

$$\frac{dc_p}{da} = \tfrac{3}{2} a^2 - 4a + 2 = 0 \Rightarrow a = \tfrac{2}{3} \Rightarrow \max\{c_p\} = \frac{16}{27}. \qquad (3.6)$$

In other words, the wind turbine can utilise up to a theoretical maximum of $16/27 \approx 59$ per cent of the kinetic energy passing through its swept rotor area. This maximum is called the *Betz Limit* and has become a virtual mantra in the wind energy community. Many would claim that, despite the simplicity and other weaknesses in the deductions, the Betz Limit of 59 per cent cannot be exceeded. Practical experience with wind turbines tends to support this claim, and modern wind turbines currently operate at efficiencies of 45–50 per cent.

However, by mounting vanes on the blade tips or other devices that concentrate the flow, it is possible to augment the efficiency. Using such vanes, one could say that the swept area is effectively increased without increasing the actual projection of the rotor contours on the vertical plane. In economic terms, such flow concentrators appear to be only partly feasible since they tend to increase loadings, and hence costs, relatively more than they increase efficiency.

**Wind turbine engineering**

As stated above, the overall momentum and energy balance considerations underlying the Betz calculations are not applicable today for practical design purposes. At the threshold of the twenty-first century, wind turbine engineering is a highly technical discipline which draws on a comprehensive framework of theories and numerical calculation methods.

Detailed discussion of modern turbine design methods is well beyond the scope of this book. However, the following pages present a brief introduction to modern design methods and considerations. Wind turbine engineering can be roughly grouped into five areas of focus.

The first area is *wind structure*. It is of vital importance in wind turbine engineering to understand the structure of the wind itself, especially issues such as the wind's turbulence and extreme values. Turbulence involves rapid changes in the wind's speed and direction, causing fatigue loads in turbines' mechanical components.

The second area, *aerodynamics*, deals with the wind's flow through the wind turbine rotor and around the blades for determination of the wind's forces on the blades and rotor. The 'work horse' for calculating aerodynamic loads and performance has been the Blade Element Momentum (BEM) method, developed by H. Glauert in the 1930s (see Glauert, 1935). The BEM method subdivides the rotor into annular sections of suitable size and forms relations between the local wind speed and blade forces. Besides the BEM model, a number of other more advanced models exist such as the generalised actuator disc model and the full three-dimensional Navier–Stokes models. As the BEM model is fairly simple and fast to use, it is well suited for design and optimisation purposes. For more specialised purposes, the BEM model can be supplemented by elements of other models such as a dynamic stall model.

Development and design of airfoils for wind turbine rotor blades (and for aircraft) was from the beginning completely empirically based. Experimental research was carried out in Göttingen (in Germany) during the First World War and later by the National Advisory Committee for Aeronautics (which subsequently evolved into NASA) in the USA. This research resulted in a theoretical framework and a number of families of airfoils as reported in the classic book by Abbott and von Doenhoff, 1959. These airfoils were widely used for wind turbine blades until the early 1990s. Today, airfoils specially designed for wind turbine blades can be developed and designed through use of advanced computer-based calculation codes. The first of these airfoil families for wind turbine blades were presented in the 1980s (Tangler and Somers, 1985; Björk, 1989).

A special issue in the field of aerodynamics is that of aeroacoustics, where aerodynamic theory is applied towards understanding and minimising the noise created by the wind's flow around rotor blades. The aeroacoustic models used today rely on a considerable number of empirical relations, whereas computational aeroacoustics (CAA) is a rather new discipline in wind turbine engineering, still at the basic research level and not yet applied in practical design.

The third area of focus involves *structural dynamics*. This area deals with wind turbines' response to aerodynamic loads from the wind. The combination of aerodynamics and structural dynamics is called aeroelastics. Comprehensive computer codes based on aeroelastic models are used for the practical determination of loads on and design of wind turbines. The code most widely used within the industry is known as FLEX4 (Øye, 1992).

The fourth area is that of *loads and safety*. The actual design loads for a wind turbine are determined by a number of critical operational modes (start-up, steady operation during high wind speeds, emergency brake and so on). For each of the operational modes the aeroelastic codes are used to simulate the loads on the turbine. When the critical design loads are determined, the allowed stress and strain in the individual components are known, and safety levels are clarified, the geometric design of the components can then be made. International standards and design guidelines play an important role in this area.

Finally, the area of *design and optimisation* comprises the 'black box' of practical wind turbine design, taking the above four areas into account. Until recently, the choice of geometric values for the rotor and final design of wind turbines was based on a combination of aeroelastic load calculation tools, guidelines and design codes, and the designers' and engineers' empirically based experiences. Today, turbines often are designed by use of numerical optimisation tools (Fuglsang and Madsen, 1996). In general an optimisation tool consists of an aerodynamic or aeroelastic code, a cost function which provides a relation between load and costs for the individual turbine components, and an optimisation algorithm. The design problem is now defined by the parameter which must be optimised, for example, minimal cost per kWh; and furthermore the design space must be bounded by constraints, for example, the size of the turbine. The advantage of the numerical optimisation tools is that a large number of design parameters can be treated, and the optimisation is performed using the whole operational interval of, for example, wind speeds.

These tools are based on two methods. For single design point methods (SD), the geometric design of a rotor is optimised for one operational situation (a certain average wind speed). For multiple design point methods (MD), the rotor can be optimised for several operational situations. Both methods are based on the BEM theory. In the late 1980s numerical methods for SD optimisation of rotors introduced a systematic parameter variation for different design parameters aiming to maximise turbines' energy production. In the early 1990s some first-generation numerical MD-based optimisation algorithms were developed, also with the aim of maximising wind turbines' production. Most recently, second-generation tools are being developed. These second-generation numerical optimisation tools are aiming directly at minimising the per-kWh costs of the turbine by linking a cost structure to the geometric design of a turbine (Fuglsang and Madsen, 1996).

To these five areas of design focus, one must of course add other major topics such as advanced electrical engineering for generator design, control engineering for design of wind turbine controllers, composite materials engineering for blade manufacturing, and so on.

## Modern wind turbine technology

Modern wind turbine technology can be classified into three main categories: large grid-connected turbines, intermediate-sized turbines in hybrid systems, and small stand-alone systems. Large grid-connected wind turbines, in the size range of 150 kW and up, account for by far the biggest market value among wind turbines. The size of commercially available grid-connected wind turbines has evolved from 20–50 kW in the early 1980s to the 500–800 kW range most common in the late 1990s. Turbines in the 1–2 MW size range have been installed as prototypes since 1995 and have been commercially available since 1997. Today grid-connected wind turbines are often placed in wind farms of 10–100 MW which are operated as a single plant. Different wind turbine design concepts are in use, the most common currently being three-bladed, stall or pitch regulated (see following section), horizontal-axis turbines operating at near-fixed rotational speed.

Intermediate-sized wind turbines in the 1–150 kW range can operate in hybrid energy systems combined with other energy sources such as diesel, small-scale hydro, photovoltaics, and/or storage systems. Intermediate turbines have significant potential for use in rural electrification. In areas with high costs of electricity and a sufficient wind resource (over 5 m/s) such wind turbine-based systems can also offer reliable and competitive solutions for applications such as water pumping, sea water desalination and so on (Hopkins, 1999). The technology for wind turbines in hybrid energy systems is ready for the market, and several potential large markets have been identified. Nevertheless, potential customers for this technology, including governments, international development banks, aid organisations, local utilities and industries, have been hesitant in installing these systems due to their lack of solid track record. Further demonstration programmes may be required to build confidence in these systems' reliability and cost-effectiveness in order to establish a solid market.

Small 'stand-alone' wind turbines of less than 1 kW for water pumping, battery charging, heating and so on represent the third turbine category. The most commercially successful in this category are very small wind turbines in the 25–150 watt range with rotor diameters of 0.5 to 1.5 metres. Such small wind turbines are widely used for battery charging at remote telecommunication stations. Yachts also often carry a very small (less than 1 kW) wind turbine for battery charging which can be used for television sets, communication systems and small refrigerators.

This book focuses on wind energy technologies used for electric applications, as these have undergone the most significant technological advances and are thought to hold the greatest promise for future applications. However, the most common technology currently in operation remains the mechanical farm wind pump. One to two million units are in regular use worldwide, with over 50 known manufacturers active in this field. The main application for mechanical farm wind pumps is for drinking water supply in rural areas, and the present annual installation of wind pumps is estimated to be on the order of 5000 to 10,000 units (Eurec-Agency, 1996).

## Principal components of the wind turbine

Wind turbines come in two broad categories: the horizontal-axis turbine whose blades appear similar to aeroplane propellers, and the vertical-axis turbine whose long curved blades are attached to the rotor tower at the top and bottom and have the appearance of an eggbeater. Vertical-axis turbines have not lived up to their early promise, and today virtually 100 per cent of existing turbines use the horizontal-axis concept. This chapter therefore focuses exclusively on horizontal-axis machines. The principal components of a modern horizontal-axis grid-connected wind turbine are illustrated in Figure 3.2 and are described below.

- **Rotor.** The rotor includes the blades and hub. The rotor can rotate either at near-fixed speed, or at variable speed, depending on the design concept. With fixed-speed operation, the rotational speed is typically 20–25 rpm for a 700 kW wind turbine, though this is dependent on design criteria. Larger turbines with longer blades have slower rotations, while small turbines with short

**Figure 3.2** Principal components of a wind turbine (pictured here is an upwind horizontal-axis wind turbine)

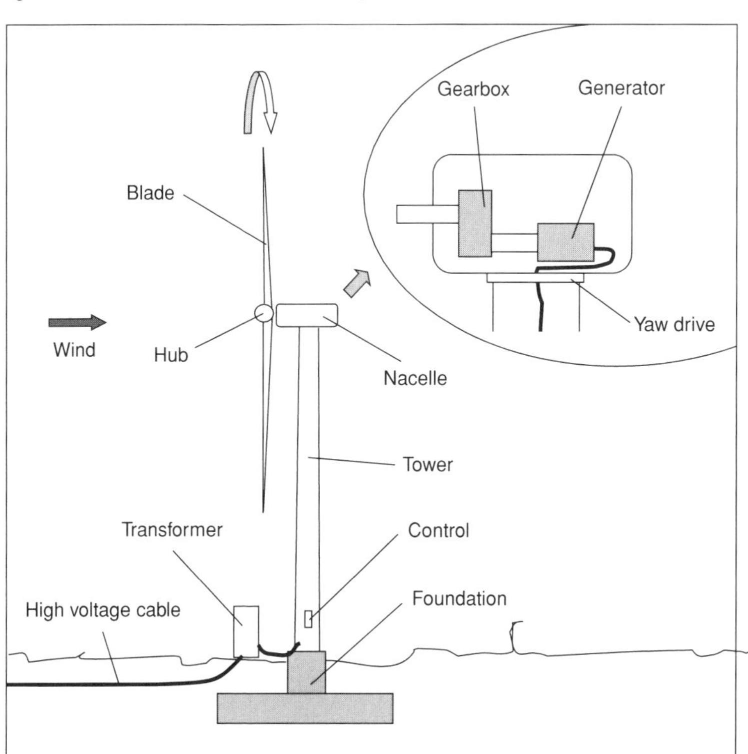


blades rotate more quickly. For a three-bladed turbine, optimum power output is typically achieved when the ratio of blade tip speed to wind speed is approximately four to one.

- **Blades**. The blades are attached to the hub. They can be attached in one of two ways: (1) in a fixed, angular position, known as stall regulation, or (2) on bearings so that the whole blade can be pitched at different angles depending on wind speed, known as pitch regulation. The cross-section, or profile, of the blade is designed to fulfil several requirements including high efficiency and good stall properties. Current wind turbines most often have three blades, but two-blade models are also common. Under stall regulation, the blade angle is set such that the blade automatically loses its lift under very high wind conditions, thus passively restricting the amount of torque on the rotor. Under pitch regulation, the angle of the blade is modified based on wind speed to provide more optimal power output over a wider range of wind speeds.

- **Hub**. The hub connects the blades to the main shaft. Hydraulic, mechanical or electrical equipment to drive the pitch setting of blades or emergency aerodynamic brakes are often mounted in the hub.

- **Nacelle**. The box-like structure located behind the rotor blades is known as the nacelle. The nacelle contains the gearbox, the generator, and various control and monitoring equipment. The nacelle is attached to the tower through the yaw drive.

- **Gearbox**. The gearbox increases the slow speed of the main shaft to a speed suitable to the generator. Thus, the speed of the rotor, which is typically well below 100 rpm, is increased up to the 1200–1800 rpm range required by the generator to produce grid-quality electricity.

- **Generator**. The generator is typically of the induction type, operating at near-fixed speed. Other generator types are being applied in newer turbine concepts as outlined in the following section.

- **Yaw drive**. The yaw drive aligns the nacelle so that the rotor axis points as accurately as possible towards the wind. Wind turbines may face either upwind or downwind. The downwind configuration is more common among small turbines and uses passive yaw control, similar to a weather vane. The upwind configuration is used in most large modern turbines and requires active yaw control, in which the yaw motor is controlled by a wind vane on top of the nacelle.

- **Tower**. The tower is typically of tubular design, particularly for large turbines. It is most often made of steel or, less frequently, of concrete. Lattice steel towers are also used but are today more common for smaller turbines.

- **Control system.** The computer-based central control panel of the wind turbine is typically mounted inside the tower (if tubular). The control system monitors gearbox and generator temperature, wind speed (if wind speed is above some set limit, the wind turbine may be stopped for safety reasons), vibration and so on. If the wind turbine is part of a wind farm, the turbine is connected to a central monitoring computer.
- **Foundation.** The tower is bolted to the foundation, typically made of concrete.
- **Transformer.** The low voltage electricity output from the generator is stepped up to grid level through the transformer. From the transformer, a high voltage cable or overhead line feeds into the main grid.

What has been described above is the 'standard concept' as of 1998. New concepts are also under development and are discussed in the following section.

## Technological trends[2]

A United Nations conference on New Sources of Energy was held in Rome in 1961. Volume 7 of the proceedings from this conference was published in 1964 and concerns wind energy (United Nations, 1964). The proceedings contain many high quality papers as well as several reporters' summations. It is astonishing how many discussion themes are the same at the end of the 1990s as they were in 1961. Comparing the state of the art in 1961 with today is quite useful and may provide some clues as to the future, in say 2030. The most visible differences between today and 30 years ago are in terms of the commercial market for wind turbines. In 1961 no commercial markets for wind turbines existed, and almost all turbines presented in Rome were prototypes or parts of research and development (R&D) programmes. Today the commercial market for wind turbines is growing very rapidly, and it is generally acknowledged that the future will bring a large, more or less stable, world market for wind turbines.

All imaginable concepts of wind turbine design were presented in Rome in 1961. No really new concepts have been developed since then, and only a few concepts enjoy a significant market share today. It is expected that the future will bring a greater variety of concepts to the market, facilitated by the growth of the overall market's volume and the available strategies for newcomers entering this market.

The prime current objective for industrial R&D in wind turbine technology is cost reduction. Wind turbine costs and, consequently, power production costs have decreased steadily since the early 1980s. Wind turbine technology's *progress ratio* (the decline in costs each time the cumulative manufactured volume doubles) has been on the order of 10–15 per cent. This development is expected to continue in the future, perhaps at a slightly slower pace.

The state-of-the-art turbine concept of the future will be a highly flexible machine. It has been generally acknowledged ever since 1961 that highly flexible turbines are theoretically the most cost-efficient. The problem is that highly flexible structures lead to high degrees of freedom in the structural dynamic design calculations. In 1961 it was not possible to model and simulate the structural behaviour of even 'conventional' wind turbines, let alone highly flexible turbines. All designs had to be verified through measurements on full-scale turbines. This has given simple, heavy concepts with a low number of degrees-of-freedom a competitive advantage. This advantage for relatively simple machines has continued until today and is likely to remain in the near future.

Recent development of fast and cheap computing technology, however, means that today engineers are increasingly able to use computer simulations of wind turbines' aeroelastic behaviour in the design process. The future will bring even faster computers, and a solid long-term R&D effort will perhaps provide a better understanding of the aeroelastic behaviour of wind turbines. At that time, the design and verification of highly flexible wind turbine concepts will become practical. Flexibility is expected to increase in future wind turbines in three primary dimensions: structural flexibility, drive-train flexibility and control flexibility.

Regarding structural flexibility, it is generally acknowledged that lightweight designs with high structural flexibility are theoretically more cost-competitive than the heavier, more rigid turbines of today. Many papers and articles have argued for such two-bladed teeter hub, downwind lightweight designs. Industrial development of lightweight turbines depends on tools for fast and reliable simulation of flexible wind turbines' aeroelastic behaviour. There is still a long way to go before such aeroelastic behaviour is fully understood and modelled. Faster computers together with the results of future R&D programmes will undoubtedly improve the tools for load simulation and determination of load cases in the future. As the tools are improved they will be utilised by the industry for increasingly flexible designs.

Higher structural flexibility also means higher drive-train flexibility (variable speed, gearless designs and so on). Such drive-train improvements will have two primary advantages: (1) increased electricity output and hence greater cost efficiency, and (2) improved power quality and grid interaction. Today several manufacturers have introduced variable-speed turbines. Introduction of further drive-train flexibility in the future will be determined by two factors: (1) the speed at which power electronics become cheaper, and (2) demands for high quality power by grid operators.

Structural and drive-train flexibility are of a physical nature. Control flexibility concerns the knowledge built into the machine. All industrial products (automobiles, watches and so on) are incorporating ever-increasing computer technology, and wind turbines will be no exception to this trend. Wind turbines in the twenty-first century will benefit from cheap and reliable computers and sensors to allow for adaptive operation. By developing flexible control systems, operating wind turbines can adapt to specific site conditions, to different safety levels, to the grid's power quality, and take into account the used lifetime of vital components, and so on.

Commercially competitive wind turbines have grown from 55 kW in the early 1980s to more than 1000 kW today. Turbines of 2 MW are already available on the market and are likely to become more competitive within the next few years. This up-scaling is expected to continue at least one step further, to a 4–6 MW turbine. In Europe such turbines will be suitable primarily in offshore wind farms. Transport and installation of very large turbines is not a problem offshore because of the availability of floating construction cranes. In other parts of the world with more land availability, 4–6 MW turbines can be placed on land, provided that logistical problems of size can be solved cost-effectively.

Investigations have indicated that wind turbines have a flat cost optimum for sizes from 500–800 kW and up. Therefore, factors other than optimising the cost of the turbine itself will determine the size of future commercial turbines. These are factors such as logistics and impact on the landscape. We expect that the size of the most competitive turbine (taking all factors into account) will differ from market to market. This provides manufacturers with an interest in developing new turbines in a variety of sizes, not merely developing ever-larger turbines, as has been the case in the 1990s.

There is no reason to believe that the introduction of highly flexible designs will happen overnight. Established wind turbine manufacturers are not likely to gamble with their expensively acquired reputations as providers of reliable turbines. But they will have an interest in remaining competitive, and the current technology will gradually adapt more flexible features. This gradualist approach has in fact been the key to the commercial success of today's largest turbine manufacturers. The order in which these incremental changes are introduced into commercial wind turbine technology will be driven by the demands of the markets, as different markets demand different designs.

Only newcomers into the wind turbine market will have an interest in introducing designs radically different from the established technology. Some newcomers are expected to make such attempts as the market volume expands and offers interesting business opportunities. On this basis, a larger variety of concepts and designs is expected in the first decade of the twenty-first century. After this, approaching the year 2030 and the maturation of wind turbine markets, only very few concepts (and very few companies) are likely to be able to attain a commanding position.

As costs have declined, other issues have also entered the agenda for industrial R&D. Wind turbine noise emissions have been lowered due to better designed blades, improved manufacturing quality of mechanical parts, and use of damper materials. Industrial designers are also increasingly involved in the design of wind turbines, leading to enhanced visual aesthetics in the landscape. This visual aspect is of considerable importance, as discussed below in Chapter 6. Turbine quality and reliability have also improved dramatically. Today, average availability[3] of modern wind turbines is on the order of 98 to 99 per cent.

## Wind energy industry

During the last 20 years the wind turbine industry has developed into a professional high-technology industry. Wind turbine manufacturing is concentrating in ever fewer companies, most of whom are European. In Table 3.1 below, the 11 largest wind turbine manufacturers are ranked by their sales ( in MW) in 1998. The figures for each manufacturer include any sales by majority owned or fully owned subsidiaries. Sales of turbines in 1998 were most likely higher than the amount actually installed. The numbers in Table 3.1 include those turbines registered as sold and manufactured but not yet installed at their destination. As can be seen from the table, the four largest companies together had a market share of nearly 70 percent in 1998. The Indian company NEPC-Micon Ltd (part-owned by NEG-Micon) has a strong position domestically and has manufactured a large portion of the turbines installed in India.

According to the Danish Wind Turbine Manufacturers Association (1999), Danish manufacturers produced 1216 MW of wind turbines in 1998 with a value of DKK7 billion (~ US$1 billion). This indicates that the global sales of the international wind turbine industry were

**Table 3.1** World's largest wind turbine manufacturers, ranked by MW sold in 1998

| *Rank* | *Manufacturer* | *Country* | *MW sold in 1998* | *MW sold Total* |
|---|---|---|---|---|
| 1 | NEG-Micon A/S | Denmark | 608 | 2 273 |
| 2 | Enron Wind Corp. | USA | 424 | 792 |
| 3 | Vestas Wind Systems A/S | Denmark | 385 | 1 878 |
| 4 | Enercon GmbH | Germany | 334 | 1 065 |
| 5 | Gamesa Eólica S.A. | Spain | 171 | 360 |
| 6 | Bonus Energy A/S | Denmark | 149 | 859 |
| 7 | Nordex Balcke–Dürr GmbH | Germany | 131 | 332 |
| 8 | MADE Energías Renovables S.A. | Spain | 105 | 232 |
| 9 | Ecotécnia | Spain | 47 | 77 |
| 10 | Mitsubishi | Japan | 38 | 279 |
| 11 | Desarollos Eólicos | Spain | 27 | 121 |
| Others | | | 113 | 2 170 |
| Total | | | 2 530 | 10 436 |

*Source*: BTM Consult (1999).

on the order of US$2 billion in 1998, which also matches estimates by the American Wind Energy Association (AWEA, 1999). The industry's annual growth rate has averaged around 30 per cent per year during the 1990s, and similar growth rates are expected in the foreseeable future, making the wind power industry one of the world's fastest growing business sectors. These growth rates have necessitated changes within the manufacturing industry.

The first change concerns the ownership and capital base of the industry. Most modern wind turbine manufacturers started as privately owned small and medium-sized enterprises, but a broader capital base is needed to fuel their current growth. Some companies have accessed capital through public issues of equity. Today, NEG-Micon A/S and Vestas Wind Systems A/S are publicly owned companies whose shares are listed on the Copenhagen Stock Exchange. Other companies have secured a larger capital base through acquisitions by larger companies. For example, Nordex Balcke-Dürr GmbH is partly owned by the large German group Deutsche Babcock GmbH. Enron Wind Corp was formed when the American energy giant Enron Corp purchased Zond Energy Systems of the USA and Tacke Wind Energie of Germany. Through this purchase (and thanks to the resurgence of the American wind market in 1998), Enron achieved 424 MW of sales in 1998, second only to NEG-Micon, compared to the 9th and 11th place rankings of Zond and Tacke in 1997 (with 38 MW and 29 MW of sales, respectively [BTM Consult, 1998a]).

The second major change in the industry is the wave of mergers which has occurred in recent years. The above-mentioned merger of Zond and Tacke into the new Enron Wind Corp is but one example of this. The Danish company NEG-Micon was formed through a merger between two Danish companies, Nordtank Energy Group A/S and Micon A/S; and NEG-Micon has since gone on to acquire three smaller wind turbine manufacturers (the Danish company Wind World, the British company Wind Energy Group and the Dutch company NedWind) and a number of vendors. This process towards fewer but larger companies is expected to continue in the years to come.

Third, as will be elaborated in the following section, globalisation has taken place in the companies' organisations for sales, manufacturing and service.

**Local manufacturing**

The main manufacturers of large wind turbines are located in Denmark, Germany, USA, India, Netherlands and Spain. The major wind turbine manufacturers have established production facilities and joint ventures in many of the world's large wind turbine markets. Several Danish wind turbine manufacturers have established production in Germany and Spain. European-designed turbines and key components are also manufactured under licence or through joint ventures in several developing countries, including India and China.

Establishment of local or regional industrial capabilities takes many forms, ranging from the purchase of imported turn-key projects to establishment of complete local wind turbine manufacturing capability. Typically, the first projects in a country are characterised by importation of the wind turbine, including the tower, as a turn-key project or a BOOT (Build, Own, Operate and Transfer) project. Usually local companies will be hired to build foundations and to establish grid connections, but the remainder of the work will be foreign-based. If a project exceeds a certain volume, simple structures such as towers might also be purchased locally.

The next step can be the establishment of a local or regional assembly plant, where all major components are provided as kits from a wind turbine manufacturer. Such assembly plants can be fully owned by the parent wind turbine manufacturer. A joint venture between a local manufacturing company and a wind turbine manufacturer is also an option, or the assembly can take place within a local order-producing manufacturer. The actual arrangement depends on the size of the market and on local industry policies.

If the local market is big enough a wider manufacturing capability is usually established in the shape of a subsidiary or joint venture in which more vital components are manufactured locally. Local networks of vendors might also be built up. This is the case in Spain, where several of the large domestic wind turbine producers are joint ventures. For example, Gamesa Eólica S.A. is a joint venture between Gamesa (a large engineering conglomerate), the state-owned SODENA (Sociedad de Desarrollo de Navarra) and the Danish wind turbine manufacturer Vestas Wind Systems A/S. The earlier-mentioned Indian company NEPC-Micon Ltd. is a joint venture between the Indian conglomerate NEPC and the Danish wind turbine manufacturer Micon.

Though no simple rules can be outlined for how and when local wind turbine manufacturing is established, three factors are important. First, the local market must be of a substantial size, for example on the order of 30 to 50 MW annually. Secondly, this market must be stable; the political context around a country's wind power programme must be stable with no rapid changes in the foreseeable future. Finally, establishment of local manufacturing can be helped by active industrial incentives. In the Spanish case some of the joint ventures are located in areas with high unemployment. Politically, creation of local jobs may be considered as important as the development of clean energy.

## Components and services

Casting a glance in one of the international wind power trade magazines or at the exhibitors' list at a wind energy conference reveals that a significant number of companies and consultancies have developed to provide services to the wind industry. Many small companies have emerged as vendors for special wind industry products, but large international engineering companies such as ABB, Siemens and SKF have also become involved.

In terms of turnover, blade manufacturers comprise a large segment of the component industry. Blade design is one of the most 'high-tech' aspects of wind energy, and several of the leading wind turbine manufacturers use specialised vendors for blades. The blades typically constitute between 15 and 20 per cent of the total cost of the wind turbine. Companies such as the Danish LM Glasfiber A/S and the Dutch merger of Aerpac and Rotorline hold a large share of the blade market.

In the 1980s a large portion of the components used in wind turbines were standard off-the-shelf components. But as the industry has matured, production volume has increased, and technological competition has called for more specialised components tailored specifically for use in the wind power sector. Today, most components such as gear-boxes and generators used in wind turbines embody a high degree of specialised experience and R&D. Manufacturers of these components have often established substantial businesses as vendors to the wind turbine industry. For example, even though there are no major wind turbine manufacturers in Finland, Finnish vendors to the international wind turbine industry had a turnover of FIM 300 million (~US$60 million) in 1997, mainly for gearboxes and induction generators (IEA, 1998).

Apart from blades, transmission systems and generators, a supply industry has also developed for components such as bearings, brakes, controllers, measurement and sensor systems, and telecommunications. Several metal fabrication companies have specialised in, for example, manufacturing of welded towers, and casting of hubs and main shafts.

Other specialised wind industry services have also emerged. In Denmark and Germany, a few banks and insurance companies have developed special competencies in insurance and financial services for wind power developers and owners. Transport companies have specialised in transporting nacelles, towers and blades from production facilities to installation sites; and crane companies have a significant business in assisting during installation of the turbines. Service and maintenance is usually carried out directly by the wind turbine manufacturer, but several independent companies also provide this service for wind turbine owners. The wind energy sector also provides opportunities for a variety of consultancies and software providers.

## Standardisation and certification

Within the framework of the EU's CENELEC (European Committe for Electrotechnical Standardization) organisation and the International Electrotechnical Commission (IEC), international standardisation of wind energy technology has taken place over the past ten years. This includes technical standards for safety and loads, quality assurance systems for wind turbine production and installation, and quality systems for certification bodies and for measuring bodies. International standardisation and certification of wind turbines are not easy tasks because different national traditions exist in this area. In the European tradition, certification is normally regulated by national (or European) authorities, and standardisation is driven by governmental institutions and research centres in co-operation with industry. In the American tradition, certification is less of a matter for the authorities, and standardisation is primarily driven by the industry itself.

It is generally acknowledged that the European-type approval and certification systems have helped the European wind turbine industry develop a reliable and competitive technology. As the industry matures in the twenty-first century, we expect that the emphasis on national standards will decline and be replaced increasingly by international standardisation, and that industry will become more actively involved in standardisation.

European certification companies such as Germanischer Lloyds and Det Norske Veritas have for a long time been active in providing type-approvals of wind turbines and certification of projects. Several other certifying bodies are active in the certification of wind turbine manufacturers' and their vendors' quality assurance systems. Recently, Underwriters Laboratories of the USA has also entered the wind power technology certification market.

## Wind energy's interactions with the electricity grid

For energy planners and power utilities, wind turbines are still considered a new technology and raise several questions regarding the turbines' behaviour in relation to the general electrical grid. Three of the primary issues of concern are wind energy's appropriate capacity credit, short-term prediction of wind resources, and power quality. Each of these issues is discussed in turn.

### Capacity credit

Capacity credit can be defined as the fraction of a power plant's rated capacity which is likely to be available at the time of peak demand (Swisher et al., 1997). Because of the intermittent nature of wind power, one cannot necessarily depend upon a wind turbine to produce electricity at any given time. As a result, it is sometimes argued that the appropriate capacity credit for wind turbines is zero, meaning that wind turbines cannot be relied upon to operate at all at the time of system peak demand. However, this is not correct.

The appropriate capacity credit of wind power can be determined by use of the 'loss-of-load-probability' (LOLP) approach, which is based on probabilistic considerations. The loss of load probability is a commonly used method for assessing the reliability of the power supply system with conventional power generation. The method calculates the probability that the generation system will not be able to meet the required electricity demand due to both forced and scheduled outages of generators in the system.

All generators in a power system, including coal, nuclear, natural gas and hydro plants, will be out of service for part of any given year due to both scheduled maintenance and occasional breakdowns. The actual availability of, for example, diesel generator sets is normally in the range of between 70 and close to 100 per cent, whereas the expected availability is usually in the range of 90–100 per cent (Hall and Blowes, 1995). There is therefore a certain probability that, because of these outages, consumer load will not be met.

When wind energy is introduced into the power supply system, the wind turbines will contribute towards meeting the system load. The question is whether wind energy can substitute for conventional generation capacity and, if so, by how much. The main problem when estimating the LOLP for a system with wind energy is how to handle the intermittent nature of wind power production. Since LOLP is a statistical framework, the natural approach is to handle wind energy production in a statistical manner. One approach is to divide a typical year into shorter time-frames having the same statistical properties. For example, one could divide the year into months and then calculate the LOLP for each hour in an average day for each month. These can then be aggregated to calculate the total LOLP. The wind power production and the load will then be described by probability distributions for these time-frames as well as the availability of the wind turbines and the conventional generation. The availability[4] of modern wind turbines is very high, being above 98 per cent.

The capacity credit of wind energy can also be defined as the amount of conventional generation that is needed in the same power system, in the absence of wind energy, to have the same LOLP. In other words, if 1 MW of conventional generation capacity were required in a given system to obtain the same LOLP as with 4 MW of wind capacity, then the wind energy's capacity credit would be $^1/_4$ or 25 per cent. The LOLP of a system can therefore be calculated with and without wind energy; and the wind energy's capacity credit will be indicated by the amount of conventional generation required in its place to obtain the same LOLP. Important issues determining the capacity credit of wind energy include the average wind speed, the variability of the wind speed, the percentage of total electricity demand that is covered by wind energy (the penetration level), the correlation between wind speed at different wind farms, and the correlation between wind speed and load.

It is generally recognised that, using the LOLP approach, at small wind energy penetration levels (up to 10 per cent of total kWh production) the capacity credit of dispersed wind turbines in a large grid is approximately equal to the wind turbines' capacity factor,[5] typically in the range of 20 to 40 per cent. The reason for this is that the fluctuations of output from the wind turbines are averaged out by the large number of wind turbines, and that because of the spatial distribution of the wind turbines more long-term variations are also eliminated. When the penetration level of wind turbines increases, the capacity credit begins to decrease as the variations in wind power output become important. However, investigations indicate that even for systems with very high wind penetration levels, the capacity credit for wind is still 50–75 per cent of the wind turbines' average output. The Republic of Cape Verde, for example, achieved a capacity credit of 75 per cent of the wind turbines' capacity factor at 25 to 50 per cent wind energy penetration (Tande and Hansen, 1996).

Though wind power production cannot be scheduled, it can be predicted, based on predictions of wind speed. Based on these predictions, the expected average as well as the expected minimum and maximum wind power production can be utilised in scheduling the dispatch of conventional generation, thereby reducing some of the stochastic nature of wind energy and increasing its value. This is elaborated upon in the following section.

### Short-term wind prediction

It is of paramount importance for utilities with high wind power penetrations to know precisely the consumption and production of both conventional and wind generated power. To take full advantage of wind farms (that is, to save maximum conventional fuel) it is also necessary to control the conventional power plant very precisely. The value of wind power to the grid can be improved by better prediction of electricity production from wind turbines.

In response to the Danish government's targets for wind energy of 10–12 per cent of total electricity supply by 2005 and 50 per cent of total electricity by the year 2030, the Danish wind energy research centre at Risø National Laboratory has placed a large research effort on this area. The approach used by Risø has been to combine numerical weather forecasting models with models that correct for very local conditions, such as wind speed-up on hills, the change in the roughness of surfaces (for example, fields, forests and water), and the shelter provided by large buildings.

The chosen numerical weather forecasting model is the HIRLAM (HIgh Resolution Limited Area Model) model of the Danish Meteorological Institute (DMI). This model runs operationally at DMI, producing forecasts twice a day for 36 hours ahead. The model domain covers Europe and beyond with a grid of around 50 km spacing. The model used to fine-tune these results to account for local conditions is the WAsP (Wind Atlas Analysis and Application Program) model developed by Risø in connection with the creation of the European Wind Atlas. The link between the HIRLAM model and the WAsP model is an equation called the geostrophic drag law, which relates the flow in the free atmosphere, where the flow is unaffected by the surface of the earth, to the flow near the earth's surface. These types of models are collectively known as Numerical Weather Prediction (NWP) models.

If it is possible to predict the wind locally, an obvious extension of the model is also to predict the electricity production from a wind farm. One of the problems in modelling a wind farm is that when the wind comes from certain directions some turbines will be sheltered by others, resulting in reduced production from these sheltered turbines. The PARK program developed at Risø can be used to quantify this effect.

The accuracy of NWP models can be tested by comparing their predictions with those of a simple persistence model, which assumes that production at any given time period is equal to the production during the previous time period. The persistence model can be represented mathematically as $P(t+1) = P(t)$.

Despite the persistence model's simplicity, it in fact predicts the weather quite well due to the characteristic time-scale of weather systems. One often experiences that weather in the afternoon is the same as it was in the morning, so for some typical weather situations the persistence model can be difficult to beat using more advanced models. Tests of the NWP-based model show that in the

**Figure 3.3** Performance of the NWP-based model compared with the persistence model for the 5.2 MW Nøjsomheds Odde wind farm

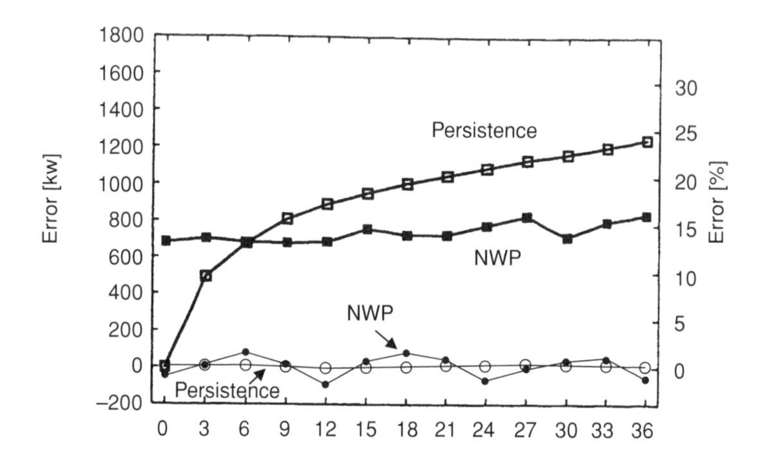


*Source*: Landberg (1999).

very short term, for forecasts of 0 to 6 hours ahead, the persistence model actually outperforms the NWP-based model.

Figure 3.3 provides a comparison between the two models for an actual 5.2 MW wind farm at Nøjsomheds Odde in Denmark (Landberg, 1999). Some of the salient features of this comparison include the following:

- The NWP prediction model outperforms the persistence model for predictions of more than 6 hours ahead. The persistence model's performance is very good for the first few hours but deteriorates rapidly thereafter.
- The mean absolute error of the NWP prediction model is around 15 per cent of the installed wind turbine capacity. Absolute error

is defined as the numerical difference between predicted and observed power production.

- The decay in performance of the NWP prediction model is very gradual. Over a 36-hour time-span the prediction error barely increases, from approximately 15 per cent of installed capacity at 6 hours ahead to 16 per cent of installed capacity at 36 hours ahead.

- The mean error of the persistence model is very small. This follows from the definition of the persistence model. Note, however, that both models have small mean errors. This is due to the fact that positive and negative errors cancel each other out, causing the mean error to be much smaller than the mean absolute error.

In Figure 3.3, the left-hand y-axis shows the total error in kW, while the right-hand axis shows the error as a percentage of total installed capacity at the wind farm. The comparisons between the NWP and persistence models are based on one year's worth of data. Similar prediction systems have been implemented and tested for wind farms in various countries and regions, including Greece, the UK and the USA (Landberg, 1999).

These results highlight the fact that wind predictions of up to 36 hours ahead are almost as accurate as predictions 6 hours ahead. This is a very important result. As discussed in Chapter 5 accurate predictions of wind speed one day in advance are crucial for wind turbines to be viable bidding into the short-term forward markets which typically characterise competitive generation markets. Theoretically, NWP models can predict as far ahead as 72 hours, provided that the model domain is global. However, Landberg and Watson (1994) estimate that realistic predictions are possible for up to 40–50 hours.

### Power quality

Wind turbines affect the quality of power in the electricity grid, and vice versa. The term 'power quality' is not always well-defined but usually covers aspects such as reactive power demand, voltage level, voltage flicker, harmonics, frequency variations and so on (Tande et al., 1996; Gerdes et al., 1997). Though concern has been expressed over wind energy's potential impact on power quality, in real-life grid operations power quality has not yet been shown to be a significant problem, even in electrical grids with a high percentage of wind power. Furthermore, technologies are available to correct wind energy's power quality impacts. Power quality issues are discussed further below.

*Voltage level*

Wind turbines produce power and thus have an influence on the voltage level in the distribution system where the wind turbines are connected. The wind turbines will influence the voltage level at the point of common coupling (PCC) where the wind turbine is connected, and from the PCC down in the distribution system. The voltage level higher up in the distribution system is also influenced, up to the point where the power system automatically regulates the voltage level.

Megavolt size (MV) substations often regulate the voltage level. In that case, the wind turbines will only influence the voltage level on the feeder to which they are connected. However, if the substation does not have such voltage regulation, then the voltage level on the MV busbar and consequently the voltage level on the parallel MV feeders will also be influenced by the wind turbines.

Wind turbines will normally increase the voltage level in the distribution system due to their active power generation. However, at the same time, the reactive power consumption of the wind turbines will decrease the voltage level. When wind turbines are installed on the grid, the voltage level is usually the designing parameter. To avoid costly reinforcements of the distribution system in that case, it can be viable to implement voltage-dependent power regulation of the wind turbines, or even voltage-dependent disconnection of the wind turbines from the grid. This will reduce the voltage level in the few periods where maximum wind energy production would have caused too high voltage levels on the grid. Statistical methods to predict the lost power have been developed and are being implemented in international standards.

*Reactive power*

Wind turbines are often constructed with an induction generator connected directly to the grid. The induction generator consumes reactive power. From the utility point of view, it is desirable to reduce the consumption of reactive power in the system, and to meet this requirement most wind turbines with induction generators also are equipped with a bank of capacitors to compensate for the reactive power consumption of the wind turbine.

As mentioned above, reduction in reactive power consumption will increase the voltage level. Therefore, high reactive power consumption can reduce wind turbines' impact on grid voltage. On the other hand, it is generally desirable to reduce reactive power consumption. The rationales for reducing reactive power are various and depend on specific grid conditions. In strong grids, the main concern is typically to reduce the losses in the system (and hence improve efficiency) by limiting the reactive power. In systems with insufficient production capacity, reactive power consumption can reduce production capacity even further. Finally, reactive power must also be taken into account when evaluating the stability of the power system.

*Voltage fluctuations*

In addition to raising the grid voltage level as discussed above, wind turbines can also cause voltage fluctuations due to their fluctuating power output (as a result of wind gusts). In other words, the current flowing from wind turbines changes with the fluctuations in power output, thereby contributing to voltage fluctuations in the grid. Voltage fluctuations in the distribution system can, depending on their frequency and amplitude, influence the power quality seen by electricity consumers and cause annoyances such as fluctuating lighting levels.

The fluctuations caused by wind turbines are of minor importance when the turbines are installed in large wind farms. In such farms, the fluctuations caused by any individual turbine are relatively small and are to some extent cancelled out by the out-of-phase fluctuations of other turbines. In addition, most wind farms are connected to the grid through dedicated lines at a higher voltage level, thus eliminating the fluctuations.

The fluctuations caused by wind turbines are of greater importance when a limited number of very large turbines are connected to rural distribution systems. With even a few turbines, however, the fluctuations of individual turbines are largely equalised by the others. On the other hand, the fluctuations from each wind turbine increase with the size of the turbine. For instance, a single 1.5 MW wind turbine will emit 2–3 times more flicker than ten 150 kW wind turbines.

*Harmonic and interharmonic emissions*

The emission of harmonics and interharmonic current from wind turbines with directly grid-connected induction generators are usually assumed negligible. However, wind turbines connected to the grid through power converters do emit harmonic or interharmonic currents and thereby contribute to voltage harmonic distortion.

The first generation of power converters was based on self-commutating semiconductors. These inverters emit harmonics of relatively low orders. To filter these requires relatively large and costly filters. Modern power converters are based on force-commutating semiconductors. Using these semiconductors with pulse width modulated (PWM) controllers can move the harmonic distortion to higher frequencies and also distribute the distortion between the harmonics as interharmonics. The filtering of these higher frequencies requires much smaller and less costly filters.

# 4
# Economics of Wind Energy

As described in Chapter 3, wind power is used in a number of different applications, including both grid-connected and stand-alone electricity production, as well as water pumping. This chapter analyses the economics of wind energy primarily in relation to grid-connected turbines, which account for the vast bulk of the market value of installed turbines.[1]

The main parameters governing wind-power economics include the following:

- investment costs, including auxiliary costs for foundation, grid connection and so on;
- operation and maintenance costs,
- electricity production/average wind speed;
- turbine lifetime;
- discount rate.

Of these, the most important parameters are the turbines' electricity production and their investment costs. As electricity production is highly dependent on wind conditions, choosing the right turbine site is critical to achieving economic viability.

The following sections outline the structure and development of land-based wind turbines' capital costs, efficiency trends, and operation and maintenance costs. Offshore turbines are gaining an increasingly important role in the overall development of wind power, and they are thus treated in detail in a separate section. The economic costs of wind power are then compared to the cost of conventional electric power. And lastly, the chapter presents a brief discussion of the economics of wind power in hybrid, stand-alone and water-pumping applications.

## Capital cost and efficiency trends

In general, two trends have dominated grid-connected wind turbine development:

(1) the average size of turbines sold on the market has increased substantially;
(2) the efficiency of production has increased steadily.

Figure 4.1 shows the average size of wind turbines sold in the Danish export market each year.[2] As illustrated in Figure 4.1 (left

**Figure 4.1** Development of average wind turbine size sold in the market (left axis), and efficiency, measured as kWh produced per $m^2$ of swept rotor area (right axis).

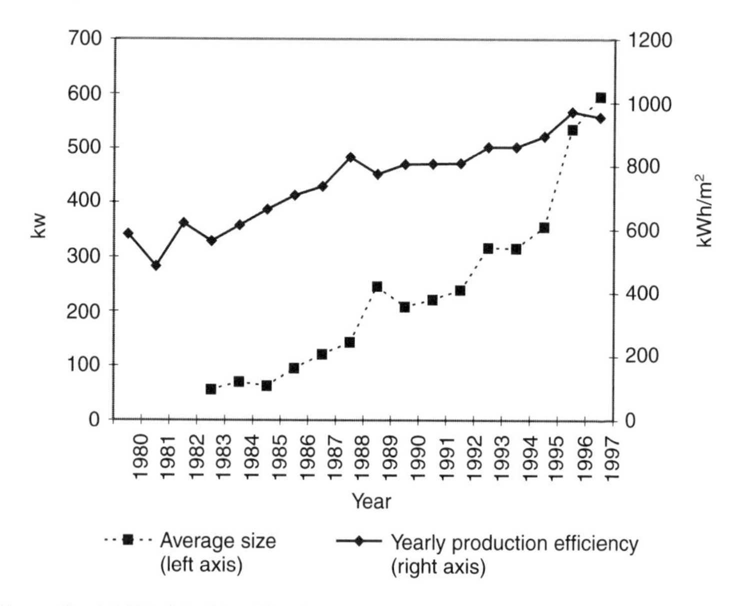


*Source*: Danish Wind Turbine Manufacturers Association (1999).

axis), the average size has increased significantly, from roughly 50kW in 1985 to 600 kW in 1997. In late 1997, the best-selling turbine had a rated capacity of 600 kW, but turbines with capacities as high as 1500 kW had already entered the market.

The development of electricity production efficiency is also shown in Figure 4.1, measured as annual energy production per swept rotor area (kWh/m$^2$ on the right axis).[3] Measured in this way, efficiency has increased by almost 3 per cent annually over the last 15 years. This improvement in efficiency is due to a combination of improved equipment efficiency, improved turbine siting and higher hub height.

Capital costs of wind energy projects are dominated by the cost of the wind turbine itself (ex works).[4] Table 4.1 shows a typical cost structure for a 600 kW turbine in Denmark. The turbine's share of total cost is approximately 80 per cent, while grid connection accounts for approximately 9 per cent and foundation for approximately 4 per cent. Other cost components, such as control systems and land, account for only minor shares of total costs.

Figure 4.2 shows changes in capital costs over the years. The data reflect turbines installed in the particular year shown. All costs are per kW of rated capacity and have been converted to 1997 prices. As shown in the figure, there has been a substantial decline in per-kW costs. From 1989 to 1996, turbine costs per kW decreased in real terms by approximately 4 per cent per annum. At the same time,

**Table 4.1** Cost structure for a 600 kW wind turbine (1997 US$)

| | *Investment (US$1000)* | *Share* (%) |
|---|---|---|
| Turbine (ex works) | 483 | 80 |
| Foundation | 23 | 4 |
| Electric installation | 9 | 2 |
| Grid connection | 53 | 9 |
| Control systems | 2 | - |
| Consultancy | 6 | 1 |
| Land | 10 | 2 |
| Financial costs | 8 | 1 |
| Road | 7 | 1 |
| **Total** | **601** | **100** |

Note: Based on Danish figures for a 600 kW turbine, using average 1997 exchange rate 1US$ = 6.608 DKK.

**Figure 4.2** Wind turbine capital costs (ex works) and other costs (US$/kW in constant 1997 $); investment costs divided by efficiency (index 1990 = 1.0)

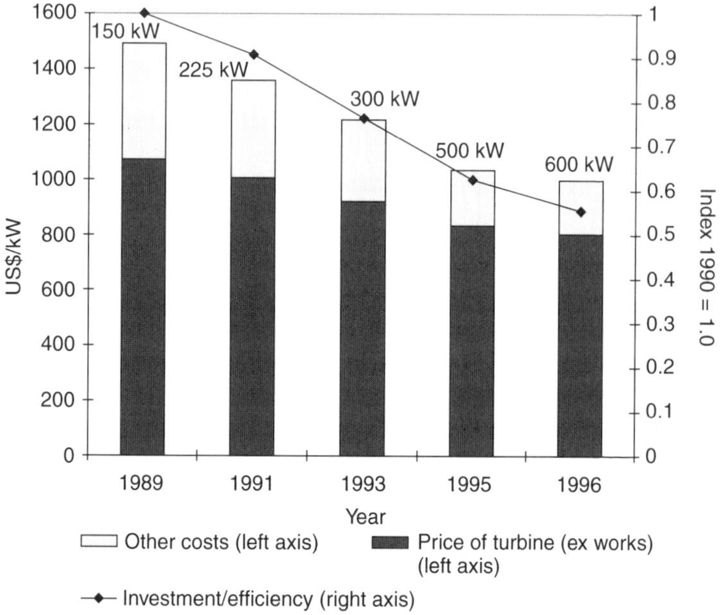

*Source*: Energistyrelsen (1994) and P. Nielsen (1997).

the share of auxiliary costs as a percentage of total costs has also decreased. In 1989 almost 29 per cent of total investment costs were related to costs other than the turbine itself. By 1996 this share had declined to approximately 20 per cent. Thus, overall investment costs per kW have declined by more than 5 per cent per year during the analysed period.

Reductions in capital costs are expected to continue for the foreseeable future. EPRI (1997), for instance, predicts that capital costs per swept area ($\$/m^2$) should decline by 23 per cent between 1997 and 2000, and by a further 10 per cent between 2000 and 2005.[5]

Combining the efficiency improvement shown in Figure 4.1 and the decline in investment costs per kW shown on the left axis of Figure 4.2, one can calculate the ratio of total investment to annual production efficiency ($/kW divided by $kWh/m^2$), shown on the right axis of Figure 4.2. This ratio provides a rough indication of total investment costs divided by annual electricity production, assuming a close relationship between turbine capacity and swept rotor area. This ratio has improved by more than 45 per cent between 1989 and 1996, or more than 8 per cent per annum in real terms. This improvement reflects not only declining turbine costs and improved efficiency, but improved turbine siting as well.[6]

Wind-energy project capital costs, as reported by the International Energy Agency (IEA, 1997a), show substantial variation between countries, owing to factors such as market structures, site characteristics and planning regulations. According to the IEA, total wind project capital costs vary between approximately US$900/kW and US$1500/kW in different countries. Caution should therefore be exercised in making cross-country cost comparisons, particularly as currency exchange rates also significantly impact on apparent costs in any given country.

## Operation and maintenance costs

Operation and maintenance (O&M) costs constitute a sizeable share of the total annual costs of a wind turbine. For a new turbine, O&M costs might have a share of approximately 10–15 per cent of total levelised cost per kWh produced, increasing to at least 20–30 per cent by the end of the turbine's lifetime. O&M costs are related to a limited number of cost components:

- insurance;
- regular maintenance;
- repair;
- spare parts;
- administration.

Some of these cost components can be estimated with relative ease. For insurance and regular maintenance, it is possible to obtain standard contracts covering a considerable portion of the wind turbine's total lifetime. On the other hand, costs for repair and related spare parts are much more difficult to predict. Although all cost components tend to increase with the age of the turbine, costs for repair and spare parts are particularly influenced by turbine age, starting low and increasing over time.

Owing to the newness of the wind energy industry, only a limited number of turbines have existed for the full expected lifetime of 20 years. For this reason, estimates of O&M costs are highly uncertain, especially around the end of turbines' lifetimes.

A small study of existing wind turbines in Denmark was conducted in an attempt to determine reasonable estimates for the development of O&M costs. For a number of different turbine sizes, average annual O&M costs were calculated for the existing turbine stock, registered in the Danish wind turbine statistics (P. Nielsen, 1997). The analysis was carried out for three successive years, 1994 to 1996. The key parameter analysed was the annual cost of O&M as a percentage of total investment costs (turbine capital cost plus installation, grid connection and control systems); and this parameter was compared to the age of the turbines to estimate the development of O&M costs over time.

Relevant O&M costs were defined to include reinvestments (for example, replacement of turbine blades or gears), if any. Owing to the industry's evolution towards larger turbines, O&M cost data for old turbines exist only for relatively small units, while data for the younger turbines are concentrated on larger units. The results are shown in Table 4.2.

**Table 4.2** Turbine age and development of O&M costs as percentage of investment costs (including costs of turbine, installation, grid connection, and control systems)

| *Turbine size* | *Sample size (varies over three sample years)* | *Average age of turbines (years)* | *Annual O&M cost as percentage of total investment costs (%)* |
|---|---|---|---|
| 55 kW | 48–57 | 10.7–12.3 | 3.1–4.5 |
| 75 kW | 16–23 | 8.9–11.0 | 2.6–3.2 |
| 95 kW | 32–50 | 7.6–9.7 | 2.7–4.5 |
| 150 kW | 69–99 | 4.5–6.4 | 2.1–2.3 |
| 225 kW | 22–25 | 3.3–4.2 | 1.8–1.9 |
| 300 kW | 5–14 | 2.5–3.9 | 0.9–1.6 |
| 500 kW | 2–34 | 1.5–3.5 | 1.0–1.9 |

*Source*: P. Nielsen (1997).

In principle the same sample of turbines should have been followed throughout the three successive sample years. However, due to the entrance of new turbines, scrapping of older turbines and general uncertainty of the statistics, the turbine sample is not constant over the three years, particularly for the larger turbines.

Care must be taken in interpreting the Table 4.2 results because the higher O&M costs of the smaller turbines may be attributed to several factors. First, as stated earlier, O&M costs increase with age, and the smallest turbines are also the oldest turbines. Secondly, the past decade has witnessed significant improvements in turbine quality, meaning that the younger (larger) turbines are better constructed and expected to have lower lifetime O&M requirements than the older smaller turbines. Thirdly, just as wind turbines exhibit economies of scale in terms of declining investment costs per kW with increasing turbine capacity, similar economies of scale may exist for O&M costs. In other words, as turbine capacity increases, O&M costs as a percentage of investment costs may naturally decrease.

For all of these reasons, one can expect that the O&M cost percentage for a 10–12-year-old 500 kW turbine will not rise to the same level as seen today for a 55 kW turbine of the same age. Most likely, the O&M cost percentage for the 55 kW turbine constitutes an upper limit to the O&M cost as a percentage of investment costs.

Figure 4.3 plots the above realised annual O&M costs along with estimated O&M cost trends over time for three sizes of turbines: 150 kW, 300 kW and 600 kW. The development of O&M costs appears, to a certain extent, to be determined by the age of the turbines. In the first few years the warranty of the turbine implies a low level of O&M expenses for the owner. After the tenth year, larger repairs and reinvestments begin to appear, and these in fact are the dominant O&M costs during the last ten years of turbine life. The estimated O&M curves in Figure 4.3 appear to be well in line with the actual O&M data points, though it must be kept in mind that the cost figures are not based on any single turbine unit, but on several turbines of different ages.

Table 4.3 summarises the results of Figure 4.3, with estimated O&M costs as a percentage of investment cost, by age and turbine size. Analyses in the remainder of this chapter assume the O&M costs shown in Table 4.3.

**Figure 4.3** Estimated and realised O&M costs over time as a percentage of investment costs, for different turbine sizes

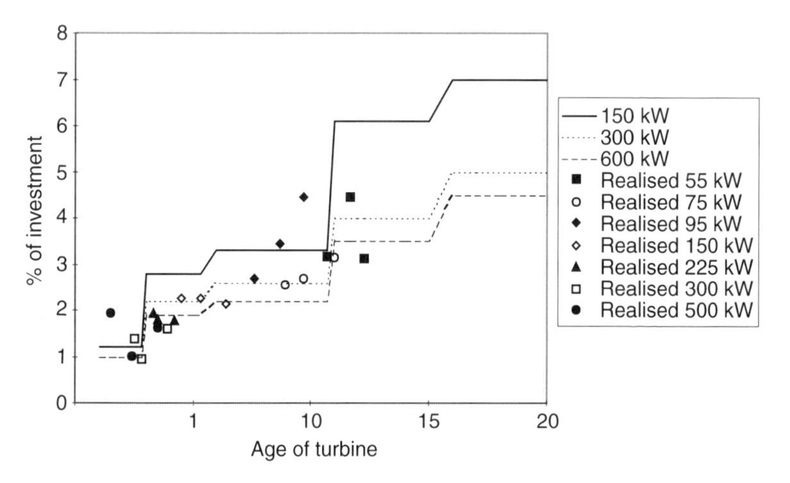


*Source*: P. Nielsen (1997); 150 kW and 300 kW turbines' curves are based on Energistyrelsen (1994), slightly adjusted; 600 kW turbine curve is estimated by the author, based on experiences from smaller turbines

**Table 4.3** Annual O&M costs as a percentage of investment cost, by age and size of turbine

| | | | *Year* | | |
|---|---|---|---|---|---|
| *Turbine size* | *1–2* | *3–5* | *6–10* | *11–15* | *16–20* |
| 150 kW | 1.2% | 2.8% | 3.3% | 6.1% | 7.0% |
| 300 kW | 1.0% | 2.2% | 2.6% | 4.0% | 5.0% |
| 600 kW | 1.0% | 1.9% | 2.2% | 3.5% | 4.5% |

## Overall cost-effectiveness

The total cost per produced kWh (unit cost) is calculated by discounting and levelising investment and O&M costs over the lifetime of the turbine, divided by the annual electricity production. The unit cost of generation is thus calculated as an average cost over the turbine's lifetime. In reality, actual costs will be lower than the calculated average at the beginning of the turbine's life, due to low O&M costs, and will increase over the period of turbine use.

Figure 4.4 shows the calculated unit cost for different sizes of turbines, based on the above-mentioned investment and O&M costs, a 20-year lifetime, and a real discount rate of 5 per cent per annum. The turbines' electricity production is estimated for roughness classes one and two, corresponding to an average wind speed of approximately 6.9 m/s and 6.3 m/s, respectively, at a height of 50 metres above ground level.

**Figure 4.4** Total wind energy costs per unit of electricity produced, by turbine size, based on hub height of 50 metres (US cents/kWh, constant 1997 prices)

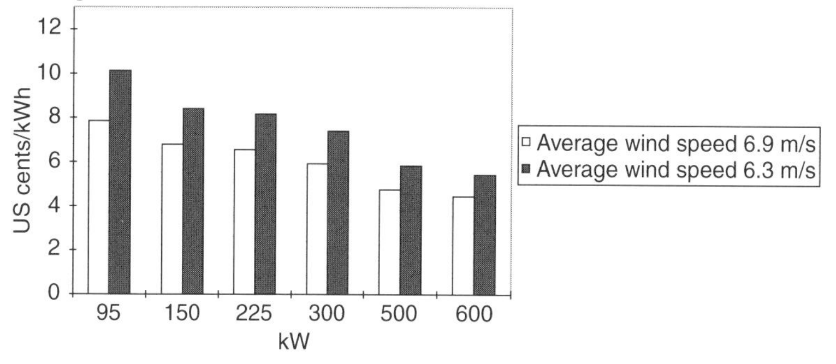


Figure 4.4 illustrates the trend towards larger turbines and greater cost-effectiveness. For a roughness class one site (6.9 m/s), for example, the average cost in 1997 US dollars has decreased from over 7.8 cents/kWh for the 95 kW turbine to under 4.5 cents/kWh for a new 600 kW machine, an improvement of almost 45 per cent over a time-span of 9–10 years.[7]

The discount rate has a significant influence on electricity production costs and hence on wind projects' financial viability. For a 600 kW turbine, changing the discount rate from 5 to 10 per cent per year (in real terms) increases the production cost by a little more than 30 per cent. Issues surrounding selection of an appropriate discount rate are complex and can differ between economic analysis and financial analysis. Discount-rate issues are discussed in detail in Chapter 5.

## Offshore wind turbines

Virtually 100 per cent of existing wind turbines are installed on land. However, locating wind turbines at sea offers several advantages (as well as disadvantages); and off shore turbines are beginning to play an increasingly important role in wind power development. Without doubt the main reason for the migration off-shore is that on-land sites are limited and that utilisation of these sites can engender opposition from the local population, as discussed in detail in Chapter 6. Land scarcity and local opposition are particularly important issues in north-western Europe, where planned intensive wind power development must coexist with high population density. As of early 1999, three offshore wind farms are in existence:

- The Netherlands recently established an offshore wind farm consisting of 19 turbines, each with a capacity of 600 kW, providing a total capacity of 11.4 MW.
- Denmark possesses two offshore wind farms, each with a total capacity of approximately 5 MW.

The latest Danish energy plan calls for more than 4000 MW of installed offshore wind turbine capacity in Denmark before the year 2030. In a recent study, the Danish utilities and the Danish Energy Agency evaluated the economic outlook for offshore wind turbines in Danish waters. The study concludes that, for offshore wind farms developed in the year 2000 with 1.5–2 MW turbines, it will be possible to achieve an electricity production cost of approximately 5.4 to 5.9 US cents per kWh (Danish Energy Agency, 1997). This is not far above the production costs encountered on land under average wind conditions.

The Tunø Knob wind farm in Denmark is typical of existing offshore wind farms and provides an illustration of the economics of offshore wind turbines. The farm consists of 10 turbines, each with a rated capacity of 500 kW, resulting in a total capacity of 5 MW. It is located 6 kilometres from the coast, at a sea depth ranging between 3.1 and 4.7 metres. Each turbine is mounted on its own separate concrete foundation, placed on the sea bottom. The turbines are connected to the high-voltage grid onshore through an underwater transmission cable. The wind farm is operated from a combined heat and power (CHP) plant located nearby, and no staff are required at the wind turbine site. Existing staff for the CHP plant perform all operating and monitoring tasks for the wind plant. Table 4.4 summarises the Tunø Knob wind farm's investment costs.

Compared to land-based turbines, the main differences in cost structure are related to three issues:

- *Foundation.* Foundations are considerably more costly for offshore turbines. The costs are related to sea depth and the selected construction principle. For a conventional turbine located on land, the foundation typically comprises approximately 4–5 per cent of total costs. In contrast, the foundation-cost share is 23 per cent for the offshore farm in Table 4.4, and thus considerably more expensive than for on-land sites. However, it must be mentioned that the foundations for this farm were developed as a pilot project and were therefore not optimised.
- *Sea transmission cables.* Connections between the turbines and the coast create additional costs compared to on-land sitings. For the wind farm considered in Table 4.4, the cost share for sea transmission cables is approximately 18 per cent.

**Table 4.4** Investment costs of an existing Danish offshore wind farm (1997 prices)

| | *Capital cost million US$* | *Share of capital cost %* |
|---|---|---|
| Turbine (ex works) | 4.8 | 40 |
| Transmission cable (sea) | | |
| to coast | 1.5 | 13 |
| between turbines | 0.6 | 5 |
| Transmission cable (land) | 0.4 | 3 |
| Electricity systems | 0.5 | 4 |
| Foundations | 2.8 | 23 |
| Operating and control systems | 0.2 | 2 |
| Environmental analysis | 1.2 | 10 |
| **Total** | **12.0** | **100** |

*Note*: Figures from the Tunø Knob wind farm, using average 1997 exchange rate: US$1 = 6.608 DKK
*Source*: Fenhann et al. (1998).

- *Environmental analysis.* A number of detailed scientific and technical investigations were carried out in relation to the Tunø Knob offshore wind project. These included investigation of the sea-bed (especially concerning debris left behind from military activities), a study to clarify the impact of the wind farms on bird life and, finally, a project to study the visual impact of the wind farm. The cost share for environmental analysis at the wind farm in Table 4.4 is 10 per cent, but a portion of these costs is related to the pilot character of this project and will probably not be repeated for future offshore wind farms.

Total electricity production from the Tunø Knob wind farm has been higher than originally expected. A net production of 15 200 MWh was generated during the first year of operation, equivalent to operation at full capacity for 3040 hours, or a capacity factor of 35 per cent. Using this production level, the investment costs outlined in Table 4.4, a real discount rate of 5 per cent, and a lifetime of 20 years,[8] total unit production costs amount to approximately 8 US cents per kWh. This estimate of production costs is subject to some uncertainty, however, due to the limited number of operating hours to-date. Again, these costs represent a pilot project; costs for future projects are expected to be significantly lower.

A number of projects have recently been undertaken in relation to minimising the cost of foundations for offshore turbines. According to Elsamprojekt (1997) the most important findings have been twofold. First, independent of the type of foundation, moving towards larger turbines will entail considerable foundation-cost advantages. Secondly, though the cost of foundations increases with water depth, this increase is less than proportional. Depending on the type of construction and the particular location, more than doubling the sea depth from 5 metres to 11 metres increases the foundation cost by only between 12 and 34 per cent.

As mentioned above, the sea transmission cable represents another important component of total costs. The closer the wind farm is located to the coast, and the higher the energy production from the wind farm, the lower will be the cost of the sea transmission cable per unit of electricity produced. Therefore, increasing the size of the wind farm will (all other things being equal) reduce the per-kWh cost of the transmission cable, except that larger wind farms must generally be located farther from land, thus re-increasing the per-kWh cable cost. The interaction of these two parameters and their impact on electricity production costs is illustrated in Figure 4.5. The figure presents an analysis of three wind farm sizes: a small farm of 7.5 MW (comparable to Tunø Knob), a medium size of approximately 30 MW, and a large farm of 100–200 MW. All wind farms are assumed to be equipped with 1.5 MW turbines.

Distance to the coast has a substantial impact on the cost of small wind farms. As shown in Figure 4.5, the production cost from a 7.5 MW wind farm increases from 4.8 to 6.7 US cents per kWh when the distance to the coast increases from 5 to 30 kilometres. Increasing the capacity of the wind farm not only significantly lowers the cost per unit of electricity produced, but also reduces the impact of distance from land. The electricity production cost for a 200 MW wind farm only increases from 4.0 to 4.3 US cents per kWh when the distance to the coast increases from 5 to 30 kilometres (Fenhann et al., 1998).

## Comparison with the cost of conventional power

The cost of conventional electricity production is determined by three components:

**Figure 4.5** Cost of offshore electricity production as a function of distance to land and capacity of the wind farm

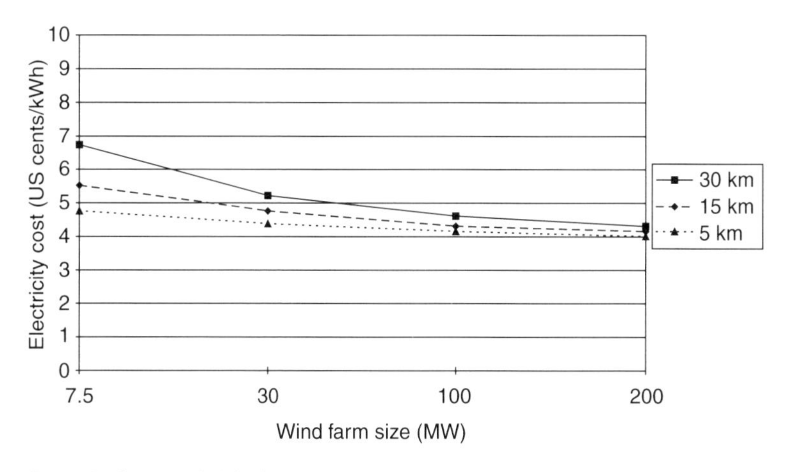


*Source*: Fenhann et al. (1998).

- fuel cost;
- operation and maintenance (O&M) costs;
- capital cost.

When conventional power is substituted by wind-generated electricity, the avoided cost depends on the degree to which wind power substitutes each of the three components. It is generally accepted that implementing wind power avoids the full fuel cost and a considerable portion of O&M costs of the displaced conventional power plant. The level of avoided capital costs depends on the extent to which wind power capacity can displace investments in new conventional power plants and is thus directly tied to wind plants' capacity credit.

The capacity credit will depend on a number of different factors, among these the level of penetration of wind power and how the wind capacity is integrated into the overall energy system. In general, for marginal levels of wind penetration, the capacity credit for wind turbines is close to the annual average capacity factor. Thus, 25 per cent is considered to be a reasonable capacity credit for wind power when the volume of wind-generated electricity is less than 10 per cent of total electricity production.[9] This capacity credit declines as the proportion of wind power in the system increases; but even at high penetrations a sizeable capacity credit is still achievable, as discussed in Chapter 3.

OECD/IEA (1998) has projected the costs of electricity generation with state-of-the-art nuclear, coal-fired and gas-fired base load power plants, given the following common assumptions:

- plants are commercially available for commissioning by the year 2005;
- costs are levelised using a 5 per cent real discount rate and a 40-year lifetime;[10]
- 75 per cent load factor;
- calculations are done in constant 1996 US$.

The OECD/IEA calculations were based on data made available by OECD member countries. Costs related to electricity production, pollution control, waste management and other environmental protection measures were included in the calculated generation costs, while general costs such as central overheads, transmission and distribution costs were excluded. Losses in transmission and distribution grids were also not taken into account.[11] Fuel price developments were projected in accordance with national assumptions.

Figures 4.6, 4.7 and 4.8 illustrate the costs of conventional electricity generation which can be avoided through wind energy, based on varying assumptions of wind energy's capacity credit. The figures are based on the above cost data from OECD/IEA (1998) for a selected number of countries and power technologies. The costs for the conventional technologies are stated in 1996 US dollars.

Figure 4.6 shows only those costs of conventional power which are avoidable through wind electricity, assuming that all conventional fuel and O&M costs are avoided but that wind power is assigned a very conservative capacity credit of 0 per cent. For example, in Spain, for each kWh of electricity generated by wind power which displaces a kWh of gas power, approximately 4.1 US cents/kWh are avoided in gas fuel and O&M costs, even if

**Figure 4.6** Projected avoided costs of conventional power compared with costs for wind-generated electricity (1996 US ¢/kWh), assuming zero capacity credit for wind power

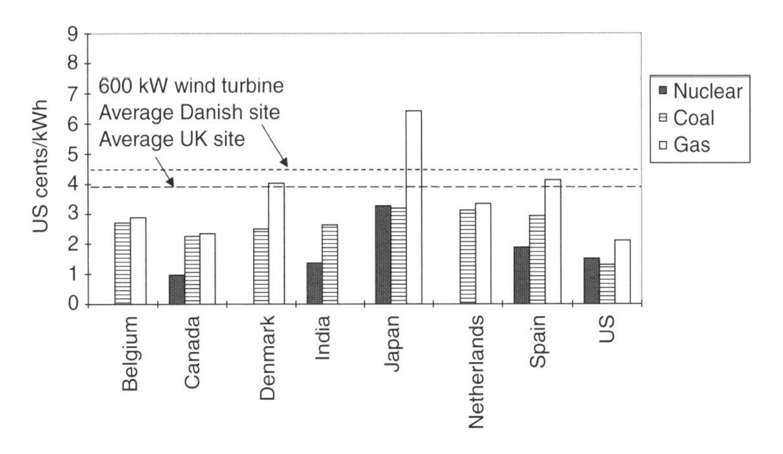


*Source*: OECD/IEA (1998).

**Figure 4.7** Projected avoided costs of conventional power compared with costs for wind-generated electricity (1996 US ¢/kWh), assuming 25 per cent capacity credit for wind power

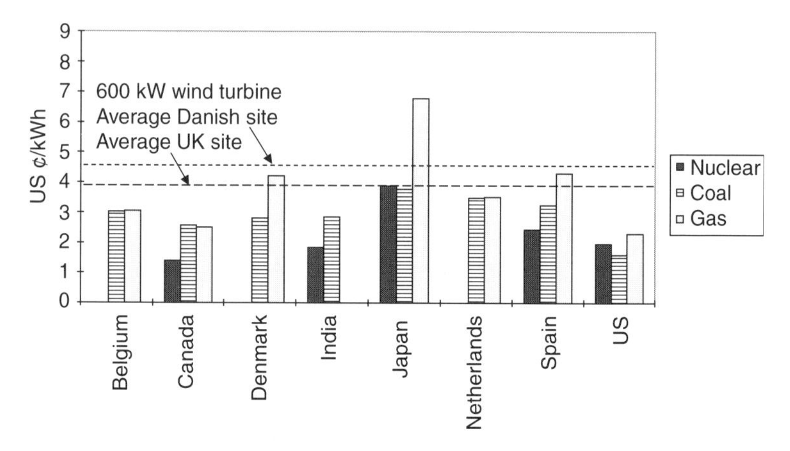


*Source*: OECD/IEA (1998).

**Figure 4.8** Projected avoided costs of conventional power compared with costs for wind-generated electricity (1996 US ¢/kWh), assuming 100 per cent capacity credit for wind power

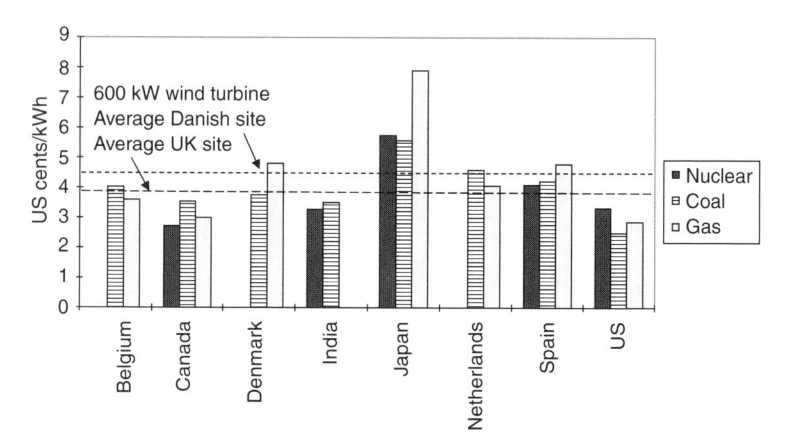


*Source*: OECD/IEA (1998).

the wind plant receives no credit for displacing any gas-plant capital costs. Therefore, if a wind turbine could be installed in Spain at the average cost of 4 US cents/kWh, this would be approximately equivalent to the avoided fuel and O&M costs of a new gas-fired power plant in Spain. For comparative purposes, the estimated total costs (including capital costs) for a 600 kW on-land turbine at average sites in Denmark and the UK are also shown (4.5 US cents and 3.9 US cents per kWh, respectively). Even under the highly conservative assumption of no capacity credit for wind energy, the 600 kW turbine is either already competitive or approaching competitiveness in terms of direct costs in a number of countries, compared to technologies based on coal and gas.

Assuming a more realistic capacity credit for wind of 25 per cent would raise the avoided costs of conventional technologies and thus improve wind's competitiveness. Because both nuclear and coal-based power costs are dominated by capital costs, assumptions about wind's capacity credit are particularly significant. The 25 per cent capacity credit assumption is illustrated in Figure 4.7.

Figure 4.8 shows the situation if one were to compare the total costs of each generating source without regard to dispatchability and capacity credit (that is, assuming a wind capacity credit of 100 per cent). The avoided costs of nuclear, coal and gas would then increase considerably, and wind would be fully competitive against conventional generating sources in many countries. The above analysis highlights the importance of capacity reliability and dispatchability. However, this importance may change in the future as electricity markets move away from centralised generation planning and towards increased competition. Much of wind energy's future competitiveness will be dependent on short-term wind predictability and the specific conditions which develop for bidding into short-term forward and spot markets. These considerations are discussed in detail in Chapter 5.

Further evidence of wind energy's improving competitiveness against conventional technologies can be seen in the 1998 results of the integrated resource-planning process of the utility Northern States Power (NSP) in the US state of Minnesota. The utility's assertion that gas-fired combined-cycle generation is the least-cost generation resource, was challenged by the public-interest group Izaak Walton League of America (IWLA). IWLA alleged that NSP used assumptions that were unduly pessimistic for wind power and overly optimistic for gas power, particularly regarding projected future natural gas prices. IWLA claimed that, using more realistic assumptions, wind energy is 32 per cent cheaper than combined-cycle gas if the wind production tax credit (see Chapter 7) is extended, and that wind is 7 per cent cheaper than gas even in the absence of the production tax credit (IWLA, 1999).

The Minnesota Department of Public Service (DPS) concurred with much of IWLA's analysis, though DPS considered wind energy's cost advantage over gas combined-cycle power to be dependent on extension of the wind production tax credit. Based on the DPS findings, in January 1999 the Minnesota Public Utilities Commission ordered NSP to proceed with 400 MW of wind power development, concluding that it was 'in the public interest under least-cost planning' (see ME3, 1998a, 1998b and 1999).

The Minnesota case is the first time that wind energy has been declared the least-cost resource in a major US proceeding. It provides some of the strongest evidence yet of wind energy's emerging economic viability as a mainstream energy source in areas of high wind availability. The case also highlights the sensitivity of gas technologies' cost-effectiveness to future natural gas prices. The low gas prices which have prevailed over the past decade have played a major role in making natural gas the new fuel of choice for electricity generation; but there is no assurance that such low prices will continue in the future.

Finally, it should be mentioned that the technology costs of coal, gas and nuclear generation shown in Figures 4.6, 4.7, and 4.8, as well as in the Minnesota proceeding, do include the costs of those pollution-control technologies required by national laws, but they *do not* include the costs of damages incurred by society for the pollution which continues to be emitted in spite of the control measures. In other words, the environmental benefits of wind energy, such as reduced human health impacts, acidification and global warming, are not captured in Figures 4.6, 4.7, and 4.8. These environmental considerations are discussed in detail in Chapter 6, which suggests that the environmental damages of fossil-fuel power plants may be worth several US cents per kWh. Including the societal benefits of reduced environmental damage in the economic calculations above would make wind energy competitive against both gas and coal-based power.

## Economics of hybrid and stand-alone wind energy systems

Extending electricity transmission and distribution grids into previously unserved areas is highly costly. CADDET (1998a) suggests distribution-line extension costs of approximately US$13 000 per mile in New Zealand, while Gipe (1993) cites utility charges of over US$30 000 per mile in France and US$50 000–60 000 per mile in California for line extensions. Higher voltage transmission-line extensions cost significantly more, often on the order of several hundred thousand US dollars per mile.

With such high line-extension costs, stand-alone wind energy systems or hybrid systems using wind and diesel and/or photovoltaics may in many cases be more cost-effective than extending the utility grid. However, small-scale wind energy applications are also expensive. While a large grid-connected turbine may have a capital cost of approximately US$1000 per kW (see Figure 4.2), small turbines in the 0.3–50 kW range may have capital costs of US$2 500–5 000 per kW (Bergey, 1998).

Nevertheless, the potential market for off-grid wind energy is very large, with more than two billion people lacking access to electricity around the world. Most but not all of them live in rural areas of developing countries. A sizeable number of people live in remote areas not served by the utility grid in developed countries as well, however. In northern California, for example, the number of stand-alone power systems grew by 29 per cent per year in the early 1990s (Gipe, 1993).

Stand-alone and hybrid systems can be cost-effective for two general situations: (1) when the utility grid is far away and not expected to be extended soon; and (2) when the utility grid is close by, but the low density and/or low electricity demand of consumers makes serving individual customers expensive. Figure 4.9 illustrates this for village electrification with solar photovoltaics (PV) in Indonesia (World Bank, 1996). In this particular case of a 3 km grid extension, when the density of households is high, it is generally cheaper to extend the grid if serving more than 40 or 50 households. If the density is less than 5 or 6 kW per square metre, however, then solar PV is generally cheaper than grid extension, even for serving several hundred customers. The situation for wind energy is expected to be broadly similar, depending on specific wind and sun conditions.

**Figure 4.9** Cost-effectiveness of solar PV homes vs. 3 km medium-voltage grid extension in Indonesia (solar PV is cheaper within shaded area)

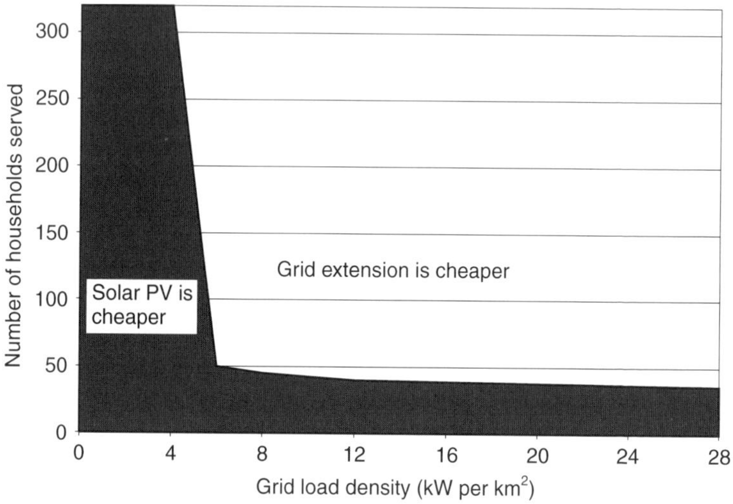

*Source*: World Bank (1996), p. 63.

In the absence of grid-based electricity, options for electricity provision include wind, PV, micro-hydro, diesel or gasoline gen-sets, and hybrid combinations of these. Generalisations of the cost-effectiveness of each option are difficult due to high site-specificity. In addition to specific wind and solar radiation conditions, cost-effectiveness is also dependent on battery price and quality, price and transportation distance for gasoline or diesel fuel, and the availability of spare parts and trained maintenance personnel.

Table 4.5 highlights the results of a comparative study of stand-alone power generation options for rural households in the Inner Mongolia Autonomous Region of China (Byrne et al., 1998). The study was conducted in four counties within Inner Mongolia, representing a variety of wind and solar radiation conditions. The study compared the cost-effectiveness of wind, PV, hybrid wind/PV, and gasoline gen-sets. All options included battery storage to allow electricity use throughout the day, though the gasoline gen-set was also analysed for non-continuous duty, without battery storage.[12]

**Table 4.5** Cost-effectiveness of stand-alone power systems in rural China

| | *System size* | *Capital cost (US$/W)* | *Output range (kWh/yr)* | *Levelised Cost (US$/kWh)* | |
|---|---|---|---|---|---|
| | | | | *Manufacturer-quoted battery life* | *Field-verified battery life* |
| Wind | 100–300 W | 1.70–2.78 | 200–640 | 0.24–0.37 | 0.50–0.63 |
| PV | 60–120 $W_p$ | 7.39–7.55 | 120–240 | 0.67–0.73 | 0.77–0.83 |
| Small hybrid | 300 W Wind + 35–60 $W_p$ PV | 2.28–3.54[a] | 400–750 | 0.31–0.46 | 0.57–0.72 |
| Large hybrid | 300 W Wind + 100-120 $W_p$ PV | 2.28–3.54[a] | 560–870 | 0.32–0.46 | 0.43–0.57 |
| Gasoline gen-set (not serving continuous duty equipment) – no battery | 450–500 W | 1.10–1.57[b] | 660–730 | 0.76–0.80 | 0.76–0.80 |
| Gasoline gen-set (serving continuous duty equipment) – with battery | 450–500 W | 1.10–1.57[b] | 480–560[c] | 1.09–1.19 | 1.16–1.27 |

*Notes*: [a] Capital costs not disclosed separately between small and large hybrid systems.
[b] Capital costs not disclosed separately between gen-sets with and without battery storage.
[c] Gasoline gen-sets with battery backup have lower annual kWh output due to battery and conversion losses.

*Source*: reprinted from Byrne et al. (1998), 'The Economics of Sustainable Energy for Rural Development: A Study of Renewable Energy in Rural China', *Energy Policy*, **26**, pp. 50–1, copyright (1997), with permission from Elsevier Science.

The results in Table 4.5 indicate that the gasoline gen-sets have the lowest capital costs, followed by wind and hybrid wind/PV. Stand-alone PV systems have the highest capital costs. On a levelised $/kWh cost basis, however, the gasoline gen-sets have the highest costs – even higher than the stand-alone PV systems. This is due to the high price of gasoline and lubricant delivered to remote communities, as well as the gen-sets' relatively high maintenance costs. Based on the manufacturer's quoted battery lifetime, the stand-alone wind systems have the lowest levelised costs, followed by hybrid wind/PV systems.

In actual field tests, however, Byrne et al. found that battery lifetimes were considerably shorter than claimed by the manufacturer, particularly for stand-alone wind systems. This in part reflects the need for improved user training for battery systems. Using field-verified battery lifetimes, the levelised cost for stand-alone wind systems increases to above that for large hybrid wind/PV systems; but, nevertheless, stand-alone wind and hybrid wind systems are the most cost-effective energy sources on a levelised basis, while gasoline gen-sets are the most expensive. In a separate analysis of rural China, Li (1996) suggests that hybrid wind/diesel systems are approximately 25 per cent cheaper per kWh than conventional diesel generation systems.

Great care must be taken in generalising the results in Table 4.5 to other regions of the world. The conditions in rural Inner Mongolia are particularly conducive to stand-alone generation with renewable energy, having very low population density, low per-capita electricity consumption, favourable wind and solar radiation conditions, local manufacture of wind turbines and PV arrays, and high cost of gasoline transportation. Nevertheless, Table 4.5 does indicate that, under favourable circumstances, stand-alone and hybrid wind energy systems can be highly cost-effective for rural electrification.

The off-grid applications of wind turbines include not only the very small (for example, < 1 kW) installations described above, but also larger turbines of up to more than 100 kW to power remote hybrid 'micro-grids' for rural communities, usually in combination with a diesel generator. Examples of such installations include a 55 kW wind turbine with a 52 kW diesel generator and short-term battery storage bank on a small Norwegian island (CADDET, 1997),

10 kW wind turbines with diesel and battery backup in northern Russia (CADDET, 1998b), and a 75 kW wind turbine with 50 $kW_p$ PV system, 60 kVA diesel engine generator, and battery storage in the Netherlands (CADDET, 1998c). Economic analysis of wind energy's micro-grid applications requires comparison with not only central grid extension but also with micro-grids based solely on diesel gen-sets. Such economic data are not readily available, and most existing installations have been implemented as one-time demonstration projects whose costs are not indicative of more widespread applications. The economically cost-effective market for wind energy in remote micro-grids is expected to be large, but significant further analysis is required.

## Economics of small-scale irrigation pumping

Large-scale electric applications have accounted for the most notable advances in wind energy and are the focus of this report. Nevertheless, the most common type of wind energy application, in terms of number of installed units, remains the small-scale mechanical wind pump for irrigation-water pumping. Such wind pumps are widely used in both developed and developing countries and can be particularly attractive when electric grid access is not readily available.

Bhatia and Swaminathan provide analyses of the economics of wind pumps for irrigation in four developing countries: Sri Lanka, Kenya, Cape Verde and Sudan (quoted in Bhatia and Pereira, 1988). Wind pumps are compared to kerosene pumpsets, diesel pumpsets, solar PV systems, or solar rankine pump systems in the various countries. Wind pumps were found to be far more cost-effective than solar systems, and in most cases they were also cheaper than kerosene and diesel pumpsets. For example, in Cape Verde, wind pumps cost between 0.080 and 0.094 US$/$m^3$ of water (1980 prices), while diesel pumpsets cost 0.082 and 0.153 US$/$m^3$, solar PV systems cost 0.49–1.73 US$/$m^3$, and solar rankine systems cost 0.90–3.28 US$/$m^3$.[13] Financial cost-effectiveness is significantly influenced by subsidies which are often available on kerosene or diesel; and the high capital cost of wind systems compared to kerosene or diesel sets also makes wind pumps' financial performance sensitive to prevailing interest rates.

CADDET (1998a) calculates the cost of water pumping with wind turbines at one site in New Zealand to be 0.37 US$/m$^3$ (based on the January 1999 exchange rate of 0.55 US$ per NZ$) at an average wind speed of 4.8 metres per second, which is considerably more expensive than the above estimates for Cape Verde, even accounting for inflation. Nevertheless, CADDET's analysis indicates that wind pumping is cheaper than electric pumping if the pumping site is more than 100 metres from the nearest electrical connection.

Gipe (1993) provides estimates of water pumping costs using both wind mechanical and wind electric systems, and using both 3 metre and 7 metre turbines. Assuming an average wind speed of 5 metres per second, Gipe's cost estimates range between 0.11 and 0.26 US$/m$^3$ for a wind electric system, and between 0.16 and 0.45 US$/m$^3$ for a wind mechanical system. These figures are roughly in line with both the Cape Verde and New Zealand results. Gipe calculates that even at low-wind sites, wind electric systems can deliver water more cheaply than diesel generators, PVs or mechanical wind pumps.

# 5
# Finance, Competition and Power Markets

The previous chapter described the changing economics of wind energy and the rapid improvement in wind energy's economic viability during the last decade. In spite of this, however, wind energy projects often continue to face obstacles obtaining finance for actual construction. This chapter therefore examines issues of finance by first outlining the differences between economic and financial viability and then describing some of the most important considerations when financing wind projects, such as cost of capital, capital structure, risk, debt-service coverage ratio, and others. The chapter then goes on to describe special considerations for financing projects in developing countries. Lastly, the chapter examines the increasingly competitive nature of power markets and how the advent of competition may affect the outlook for wind power development.

## Economic vs financial viability

It is important to clarify the differences between economic and financial viability. More specifically, we must understand why wind projects continue to face financing difficulties in spite of their improved economic viability. Three primary differences commonly distinguish financial analysis from economic analysis (see Layard and Glaister, 1996):

- consideration of investor welfare vs. societal welfare;
- consideration of private costs vs. public costs and the handling of transfer payments such as taxes and subsidies;

- differences in the discount rate used to value capital flows.

These differences can mean that a project which is economically viable from a societal perspective may not be financially viable from an investor's perspective, or vice versa. Each of these three differences is described further below.

**Investor welfare vs. societal welfare**

Financial analysis examines a project's costs and benefits purely in terms of those impacts which directly affect the project investors' welfare. Thus, private investments are typically made with the goal of maximising the net present value of private financial flows or maximising investors' convenience, leisure and so on. Economic analysis typically takes a broader societal perspective, in which case significantly different conclusions may be reached.

For example, whether an individual buys an automobile or takes the public train will be decided based on the two modes' relative costs and convenience. The ever-increasing use of private automobiles suggests that, from a private perspective, the automobile offers a sound financial investment compared to public rail transit. However, road transport may impose greater negative societal impacts than rail in terms of increased pollution, accidents, urban sprawl and destruction of natural landscapes, national balance of payment deficits from increased dependence on imported oil, and so on. The concentrated private benefits of automobiles may thus be achieved at a significant but diffuse public cost. Such costs and benefits which accrue to society at large rather than to project investors, are known in the economic literature as 'externalities' and should be included in a complete economic analysis. When societies build capital-intensive public transit systems in spite of their apparent lack of financial viability, it is generally through either implicit or explicit recognition of these additional societal benefits which they provide.

In practice, however, externalities are often either analysed separately or even, unfortunately, ignored during economic analysis because of difficulties in their quantification. In fact, the previous chapter's discussion on wind energy economics did not include an analysis of externalities. As such, Chapter 4's economic analysis is incomplete and includes only those costs for which established prices exist. If wind energy's added benefits of reduced pollution were included, wind energy would be even more economically viable than shown in Chapter 4. It is precisely this recognition of unquantified environmental benefits which has prompted many societies to invest in wind energy to date even when straight economic analysis has not supported the investment. Though Chapter 4 has concentrated solely on actually quantified economic inputs, consideration of environmental externalities is taken up separately in Chapter 6.

**Private costs vs. public costs**

The difference between private and public costs is manifested in several forms. These include:

- externalities;
- market prices vs. shadow prices;
- transfer payments including taxes and subsidies;
- cash flow constraints and access to capital.

The issue of externalities is related to the issue of investor vs. societal welfare discussed above. If the value of externalities such as pollution were reflected in market prices, then private financial costs would more accurately reflect public economic costs, and those financial decisions which maximise investor welfare would be more likely to maximise societal welfare as well. As things currently stand, the lack of consideration of externalities may mean that an energy project with the lowest private costs may not have the lowest public costs, resulting in suboptimal development for society.

The second difference between public and private costs is the discrepancy between market prices and shadow prices. Financial investment decisions are based on market prices; but economic analysis is correctly done using the economic concept of shadow prices, defined as the 'accounting price that reflects an estimate of the opportunity cost of providing or eliminating an additional unit of the good' (Zerbe and Dively, 1994). In a perfectly functioning market, market prices would in fact reflect opportunity costs, and there would thus be no difference between market and shadow prices. In well-developed economies, it may be generally acceptable to assume equivalence between market and shadow prices, but this is not always the case. Developing economies often contain greater economic distortions, including price controls, import restrictions and so on, in which case market and shadow prices may diverge substantially. For example, if importation of certain equipment is restricted in a country, then the financial price paid for the equipment within the country would be higher than the shadow price available in world markets. Thus, if a wind turbine is imported and therefore subject to import controls, its market or financial price would be higher than its shadow or economic cost. Or in labour markets containing high structural unemployment, if a project hires workers who were previously unemployed, then the appropriate financial cost of this labour would be the market wage; but the appropriate economic cost (shadow price) would be the labour's opportunity cost, represented by the value of the workers' leisure time lost by accepting employment.

The third and perhaps most important difference between financial (private) costs and economic (public) costs is related to the above concept of shadow prices and involves the treatment of transfer payments including taxes and subsidies. A tax paid by an individual clearly has a financial impact on that individual; but in economic terms this tax may be considered a transfer payment from the individual to the state which does not consume actual resources. If the tax does not represent a true resource cost, then it is not considered an economic cost even though it is a financial cost. Similarly with subsidies, if the state provides a subsidy to investors to build wind energy projects, the subsidy would always be considered in the financial analysis but not in the economic analysis. In other words, taxes and subsidies are commonly assumed to have a shadow price of zero.

The fourth area of distinction between public and private costs is regarding access to capital, including issues of debt service and cash flow constraints. Economic analysis will often assume a perfect capital market in which there are no constraints on the availability of capital. As a result, as long as a project has a positive net present value (NPV) at the appropriate discount rate, the economic analysis is not troubled by the prospect of negative cash flows in any particular year. From a private financial perspective, cash flow constraints are a very real concern and can substantially affect project costs. As discussed later in this chapter, the debt-service coverage ratio

(DSCR) is a key criterion used by lenders to energy projects to ensure that the borrower will always have sufficient cash flow in each year to meet debt service obligations. A project would typically not be acceptable to a lender if the project faces one or more years in which its cash flow is less than its debt service needs, even if the overall project NPV is positive. Thus, a project which appears economically viable might not receive financing due to cash flow constraints and limited capital availability.

### Discount rate

Financial flows that occur over a period of time must be discounted to reflect the time value of money. For a private investor, the discount rate to be used in evaluating a project will be the opportunity cost of capital for an alternative project of similar risk. As a project's risk level increases, its discount rate also increases. If an investor must pay a rate of return of 10 per cent to finance a project, for example, then this would represent the project's appropriate opportunity-cost discount rate given its risk level. If the risk-free interest rate for government treasury bills were 3 per cent, then the difference of 7 per cent between the 10 per cent project discount rate and the 3 per cent risk-free rate would represent the risk premium assigned by the financial markets for the private investor's project. The financial markets thus provide a clear indication of the appropriate discount rate to be used in financial analyses.

For economic analysis, the appropriate discount rate may be less clear. If economic analysis is assumed to represent the public perspective, public perceptions of risk and time preference may differ significantly from private perceptions. Substantial economic literature exists regarding the appropriate discount rate for public investment decisions (see, for example, Arrow and Lind, 1970; Kula, 1987). If one uses the social time preference rate as the appropriate discount rate for economic analysis of public investments, one typically obtains a real discount rate in the 2–5 per cent range (Kula, 1987; Sharma et al., 1991). This is also roughly compatible with the risk-free government borrowing rate employed by some analysts.

In terms of wind energy and its potential for combating long-term environmental concerns such as global warming, UNEP (1998) suggests a real societal discount rate of 3 per cent for analysing climate change mitigation projects. In contrast, the nominal rate of return actually demanded by financial investors in wind energy projects may be over 15 per cent. With this very large discrepancy between social discount rates used in economic analysis and financial opportunity-cost discount rates used in financial analysis, the two analyses can indicate significant differences in viability, especially considering the capital-intensive nature of wind energy projects. Thus, projects which appear to be viable from a societal perspective may not find investors who are willing to undertake them.

In conclusion, several factors can lead to discrepancies between the results of financial and economic analyses. Consideration of tax credits and subsidies will serve to improve wind projects' financial viability compared to their economic viability. Meanwhile, consideration of environmental externalities, debt-service coverage requirements and, most importantly, differences in discount rates will serve to improve wind projects' economic viability over their financial viability.

## Financing wind power projects

This section provides a brief primer on wind power finance. Having outlined the factors which cause economic and financial viability to diverge, we now consider in more detail the most important financing factors affecting wind energy projects. Before continuing, however, a brief explanation is warranted regarding the differences between *corporate finance* and *project finance*. Under corporate finance, a company's entire assets and revenue stream serve as collateral for financing a project. Thus, if a large electric utility company obtains a loan using corporate finance to build a wind power facility, the utility's entire asset-base of generators, transmission lines, distribution systems and the resulting revenue stream from customers will all be considered by the lender when determining the terms of the loan.

Project finance, by contrast, treats a project (such as a power plant) as a stand-alone legal entity with no recourse (or limited recourse) to the parent company. In other words, only the project's own revenue stream can be used to pay the project's debt obligations. Thus, if a utility built a wind power facility using project finance, the lender could not tap the utility's large asset-base to recover the loan in case the wind project failed to perform as expected. As a result, the non-recourse nature of project finance places greater risk on the providers of capital, meaning that the cost of capital using project finance is typically higher than when using corporate finance. The risk premium charged in project finance is reduced to some degree, however, by the use of extensive loan covenants by project finance lenders. Such covenants place significant restrictions on the nature of activities which may be pursued by project owners, thus limiting the potential for misuse of funds.

For various reasons, large utility companies normally use corporate finance for their investments, while smaller independent power producers (IPPs, also known as 'non-utility generators', or NUGs) typically use project finance.[1] The history of wind power projects has been dominated by IPP-style development rather than by utilities. As a result, project-finance structures have been more prominent; and the following discussion generally assumes the use of project finance rather than corporate finance. Nevertheless, the general principles are equally applicable for corporate finance as well. For a detailed analysis of the differences between corporate, project and public finance for wind projects, please refer to Wiser and Kahn (1996).

**Risk**

First and foremost to understanding the nature of finance in general and wind power finance in particular is the concept of risk. Investors are risk-averse, and the rate of return demanded by investors rises in direct relation to the level of risk they face. What exactly is risk? In investment terms, risk is defined as the variability of returns on an investment, represented by the standard deviation of the return (see Francis, 1993; Brealey and Myers, 1996). Thus, suppose investment 'A' provides an average annual return of 10 per cent, but actual returns in any given year can vary between 5 per cent and 15 per cent, while investment 'B' provides a certain annual return of exactly 10 per cent. Both A and B provide the same average return, but A is more risky than B. As investors are risk-averse and prefer certainty of returns, they would choose investment B over A.

Because power-generation projects require high up-front capital investments to be recovered over many years, they are particularly sensitive to risk. Risks arise in energy projects due to uncertainties in a wide variety of factors, including technology performance, fuel cost, customer energy demand, customer financial viability, political instability and so on. With the presence of each risk factor, the rate of return demanded by investors increases accordingly. Financiers will require that project risks be mitigated to the greatest extent possible through firm contracts and guarantees with equipment suppliers, fuel suppliers, power purchasers, governments and so on. If the unmitigated risks are deemed to be too high, projects may not be able to attract financing at any price.

Standard & Poor's, a US investment-rating agency, assesses private power projects' creditworthiness based on seven primary factors (Chew, 1995a, b, c). Chew outlines these risk considerations as follows:

*Output sales contracts*

- How is the power purchase contract structured to ensure an adequate revenue stream?
- What is the relative reliance on capacity payments vs. energy payments?
- For capacity payments, how high must plant availability be? Does the capacity payment cover debt service and fixed O&M?
- How are energy payments related to fluctuating fuel prices and variable O&M costs?
- What regulatory-out clauses are included in the contract which allow the government to disallow certain energy or capacity payments?
- For cogeneration projects, how is the steam purchase contract structured?

*Power costs*

- How expensive are the project's power costs in competing with other power plants? The cost charged by the seller to the utility must be low enough for the utility to want to dispatch the plant, but high enough to be profitable for the seller. The project's power costs will be influenced by project technology efficiency, site acquisition costs, plant proximity to fuel supply, and market fluctuations in financing, fuel and operating costs.

*Fuel risk*

- How well are fuel costs matched to electricity sales prices?
- Can increases in fuel costs be passed on to the electricity purchaser?

- Is the project over-relying on spot-market fuel purchases compared to longer-term contracts?

*Project structure*

- What percentage of the project's funding is through equity vs. debt? With non-recourse project financing, sufficient equity capital is necessary to keep owners committed to project viability.
- Does the project structure sufficiently prevent withdrawal of ownership equity and limit distributions to owners until proper cash flow and capital reserve requirements are satisfied?
- Does the project structure sufficiently restrict diversion of project funds into other assets and preclude sale of assets or ownership interests subject to bondholder approval?
- Does sufficient liquidity exist to cover temporary project difficulties? This is particularly important for higher-risk technologies, and at least 6 months of debt service and O&M costs should be held in reserves.
- Does project insurance sufficiently cover replacement value of all operating equipment and provide business interruption insurance in case of a catastrophic event? For less proven technologies, will other operational difficulties be covered by insurance? Does the insurer have an investment-grade credit rating?

*Technology risk*

- What is the level of construction risk? In other words, what is the chance that the project will not reach acceptance as scheduled and budgeted?
- Does the project involve a relatively simple technology and design, or is a more complex design being employed? What level of equipment performance warranties are provided?
- What is the construction capability and financial strength of project contractors?
- What level of guaranty is provided by contractors or third parties to fulfil all construction requirements? Are construction contract damage payments sufficient to cover project debt service requirements?
- What level of operating risk is involved? Will units provide the level of thermal efficiency required to reach project financial performance goals? Are long-term O&M expenses likely to be in line

with those projected? Is unit availability likely to be sufficient to meet performance requirements?
- What is the level of operator experience? If operation is provided by a third-party service provider, are their expenses contractually controlled and sufficient performance incentives provided?

*Power purchaser's credit strength*

- How likely is the power purchaser to meet its contract obligations to purchase sufficient power to keep the project financially viable? What is the power purchaser's bond rating?

*Projected financial results*

- What is the likelihood that cash flow will be sufficient to meet fixed charges (principal, interest, lease obligations), non-recurring capital requirements and O&M?
- What level of cash flow is dependent on non-certain sources such as non-contract spot power sales?
- Are floating-rate debt and foreign currency debt fully hedged against harmful fluctuations?

In addition, the International Finance Corporation (the private-sector arm of the World Bank Group) lists environmental risks as another significant project risk factor (Bond and Carter, 1995). These risks include: (1) hazards such as fires or explosions, (2) violation of environmental regulations such as emission standards, (3) site contamination, and (4) special concerns such as resettlement of indigenous populations.

In light of the above risk factors, wind energy projects enjoy advantages compared to conventional fossil fuel technologies in terms of fuel risk and environmental risk. Regarding technology risk, wind energy's status is less clear. Wind turbine technology is now mature and highly reliable, achieving availabilities of 98–99 per cent (see Chapter 3). And given that the most critical components come fully assembled from the factory, field construction is simple and relatively risk-free. In reality, therefore, technology risk associated with wind energy may now be as low or possibly even lower than that of more established fossil fuel technologies which require significant customised on-site construction.

On the other hand, the financial community itself is generally less familiar with wind technology and largely continues to view it as an 'unconventional' and hence risky new technology. This perception is strongly influenced by the history of wind power development in the USA in the early 1980s, when actual performance of many wind turbines was far below that promised by project developers (see Cox et al., 1991). Brown and Yuen (1994) emphasise this institutional memory factor as a particularly important barrier to obtaining finance for wind projects. Perceptions also continue to be coloured by the propensity to lump all wind technologies into one uniform category. Thus, the failure in the 1990s of the much-vaunted Kenetech (the most prominent US wind turbine manufacturer through the 1980s and early 1990s) variable-speed wind turbine to perform as expected, combined with Kenetech's bankruptcy in the mid-1990s, is likely to cause many investors (particularly American) to shy away from all wind energy technologies in spite of the fact that Danish turbines, for example, have been operating reliably for over a decade.

Similarly, under project structure risks, Standard & Poor's indicates that less proven technologies will require higher liquid capital reserves and insurance coverage, thus further raising costs. The perceptions of technological risk by the financial community must therefore be addressed in light of wind turbines' actual technological reliability which by the late 1990s has been well established.

The most important risk factor facing wind energy projects may be in the area of output sales contracts. Utilities purchasing power from IPPs typically provide separate payments for energy and capacity. Energy payments are made for each kWh generated, while capacity payments are made per kW of firm capacity available at the time of the utility's need. Given the variable nature of wind resources, utilities may not be willing to make capacity payments to wind power generators.[2] Thus, even if wind power costs are economically competitive with conventional technologies, wind power's lack of dispatchability may result in less favourable power purchase contracts, making wind power less financially viable than conventional power plants.

Some of this risk may be mitigated to some degree with the emergence of day-ahead and hour-ahead forward and spot markets in competitive power markets, as discussed later in this chapter. With accurate short-term wind prediction now increasingly possible wind power generators may be able to bid reliable power into short term markets. On the other hand, as outlined by Standard & Poor's under their 'projected financial results' risk category, over-reliance on non-certain spot power sales is also disfavoured by investors.

Thus, the single most important factor affecting the risk and hence 'financeability' of wind energy projects is the availability of an acceptable long-term power purchase contract. This has been a critical factor not just for wind power, in fact, but for all forms of IPP-developed conventional power plants as well. More recently, however, IPPs have begun constructing high-efficiency gas-fired power plants without complete long-term power purchase contracts, known as merchant plants. This has been possible under the assumption that these new merchant plants would be competitive under virtually all circumstances and would thus be financially viable even without contractually guaranteed long-term sales. For wind power, lack of dispatchability means that, despite low overall costs, merchant wind plants are not likely to be financially feasible in the near future. The importance of power purchase contracts is further discussed later in this chapter as well as in Chapter 7.

The variability of wind resources also leads to another type of risk. Though wind energy avoids fuel price risk because its 'fuel' is free, it faces instead a resource variability risk. In spite of the increased availability of accurate average wind resource assessment and short-term wind prediction techniques, there will always continue to exist a risk of climatic variation from year to year, resulting in fluctuating annual wind resources, just as hydroelectric power is vulnerable to annual changes in river flow. Such annual fluctuation means that wind power projects may face cash flow shortages in the event of a 'low-wind' year, such as occurred at California's Altamont Pass in 1994, for example (Brown and Yuen, 1994). In Denmark, an analysis of wind data over 20 years indicates a standard deviation in annual average wind speed of +/- 10 per cent and a maximum deviation of +/– 20 per cent from the long-term average.

These various risks can in turn result in a higher cost of both equity and debt capital and/or a higher required debt-service coverage ratio (DSCR), which also leads to a higher overall cost of capital. Kahn (1995) provides an overall comparison of financing costs between wind and fossil-fuel power plants. Kahn estimated the cost of equity capital for two IPPs engaged in fossil fuel-based power plant development at 8.37 per cent and 11.14 per cent, respectively. In comparison, his estimate of the cost of equity capital for a wind power development company was 17.36 per cent, indicating a significantly higher level of perceived overall risk for the wind energy company.

Kahn's estimates of equity capital cost were derived using the capital asset pricing model (CAPM), a model widely used in the financial industry, which assumes that a company's cost of capital varies in direct proportion to its level of 'market' or 'undiversifiable' risk (see Brealey and Myers, 1996). The rate of return expected by investors is determined based on the guaranteed rate of return provided by a riskless asset (like treasury bills), the expected rate of return on the overall stock market and the riskiness of the individual company being analysed compared to the riskiness of the market as a whole. A company's level of market risk is represented by its *beta value*, in which the overall market has a beta of 1.0. If a company has a beta greater than 1.0, it is considered an aggressive stock with higher riskiness than the stock market as a whole (and hence requiring a higher rate of return). Conversely, a company with a beta of less than 1.0 indicates a defensive stock with relatively small price movements. Kahn estimated the beta values of the two fossil-fuelled IPP companies to be 0.52 and 0.85, respectively, indicating risk levels for these two companies well below the level of risk in the overall stock market itself. The wind power company, on the other hand, had an estimated beta value of 1.59, indicating a level of risk much higher than that of the other two IPPs as well as the overall market itself. This high level of perceived risk by investors results in a considerably increased cost of capital, which translates into an increased per-kWh cost of wind energy.

However, Kahn's cost-of-capital calculations were based entirely on companies in the USA, where wind power has had a notably volatile and troubled history (see Chapter 7). The level of riskiness perceived in other countries can be substantially different, as discussed in the following section; and even in the USA, perceptions of high risk for wind energy projects have lessened somewhat in the late 1990s, compared to the time of Kahn's analysis.

### Capital structure

Risk is not only the most important factor influencing the cost and financeability of wind energy projects, but its influence can be observed through a variety of forms, such as capital structure, by which we refer to the relative reliance on debt vs. equity capital. The characteristics of debt and equity finance are briefly described below.

Debt generally consists of loans borrowed from banks and bonds issued on the capital markets. The main characteristic of debt is that the borrower must repay the interest and principal according to a pre-set schedule, regardless of the borrower's profitability or ability to repay. Because a lender's only profit will come from the interest payments received,[3] the lender's profit will not increase if the borrower invests in risky projects which yield high returns. Because the lender's up-side earning potential is limited, the lender will prefer conservative investments with lower risk which provide a higher likelihood of successful loan repayment.

Equity represents ownership in the firm and consists of retained profits and shares issued either privately or through a stock market. Equity investors in a company are typically rewarded through dividend payments and through appreciation of the shares' market value. Equity holders can only be paid after all of the company's debt obligations have been met. However, unlike debt holders, equity holders have unlimited up-side earning potential from profitable investments.

Hybrid forms of capital such as subordinated debt and preferred equity are also used, which exhibit some characteristics of both debt and equity (see Kahn et al., 1992; Brealey and Myers, 1996). Such hybrid instruments typically account for only a small proportion of capital raised, however, and will be ignored in this discussion for the sake of simplicity.

Because debt holders (creditors) have prior claim on a company's revenues over equity investors, creditors face a lower level of risk than equity investors. In return for accepting higher risk, equity investors demand a correspondingly higher rate of return. For this reason, it is cheaper for a company to raise capital through debt than through equity. A company or project's overall cost of capital is determined using the following basic formula:

$$WACC = (W_d \cdot C_d) + (W_e \cdot C_e), \qquad (5.1)$$

where:
$WACC$ = weighted average cost of capital

$W_d$ = weighting or proportion of debt
$C_d$ = cost of debt
$W_e$ = weighting or proportion of equity
$C_e$ = cost of equity.

For example, if debt costs 8 per cent/yr and equity costs 12 per cent/yr and a company raises 50 per cent of its capital through debt and 50 per cent through equity, then its *WACC* would be (0.50 · 8%) + (0.50 · 12%) = 10.0%/yr. As debt is cheaper than equity, the weighted average cost of capital in the above formula appears to decrease as the relative proportion of debt increases. Because the cost of capital has an inverse effect on a company or project's profitability, the formula suggests that profitability is maximised by maximising the proportion of debt.

However, financial theory suggests that, in a perfect capital market with no taxes or bankruptcy costs, capital structure should have no effect on a company's cost of capital or profitability (Modigliani and Miller, 1958). This is because, as the proportion of debt increases in a company's capital structure, the level of risk for equity owners also increases since equity owners can only receive profits after all debt obligations have been paid. As a result, the rate of return demanded by equity investors (that is, the cost of equity capital) increases as the proportion of equity in the capital structure falls. Thus in our example from the previous paragraph, if our hypothetical company raises its proportion of debt (costing 8%/yr) from 50 per cent to 80 per cent, then the cost of equity might rise from 12%/yr to 18%/yr as the proportion of equity falls from 50 per cent to 20 per cent. The resulting WACC then would be (0.80 · 8%) + (0.20 · 18%) = 10.0%/yr, no different than its previous level.

In reality, however, capital structure does in fact affect the cost of capital, largely due to the presence of taxes. In many countries, interest payments on debt can be deducted as an expense from companies' taxable profits. As a result, the earlier WACC formula can be modified to calculate the after-tax weighted average cost of capital as:

$$WACC^* = (1 - \tau)\,(W_d \cdot C_d) + (W_e \cdot C_e), \qquad (5.2)$$

where:
$WACC^*$ = after-tax weighted average cost of capital

$\tau$ = marginal tax rate.

The after-tax weighted average cost of capital formula shows a clear advantage to maximising the proportion of debt in the capital structure. This advantage may be tempered by the fact that, for investors, capital gains on equity are often taxed at a lower rate than interest income on debt (see Brealey and Myers, 1996). Nevertheless, the overall result of taxation has typically been to strongly favour the use of debt over equity in financing energy projects. In analysing twelve project-financed fossil fuel-based IPP projects, Kahn et al. (1992) found that only one project had a senior debt ratio[4] of less than 75 per cent.

The preference for debt may also be partially explained by another factor. In spite of the earlier-cited financial theory that the cost of equity capital should rise in direct relation to the proportion of debt in the capital structure, Wiser and Kahn (1996) suggest that this rise in the cost of equity due to debt leveraging has been relatively minimal in the USA.

One possible explanation for this may be that the nature of energy project finance limits the true risks borne by equity investors. Energy projects financed through project finance have typically been characterised by two fundamental conditions: (1) the project must obtain a power purchase contract from the utility, guaranteeing that the utility will purchase the electricity generated by the project; and (2) all significant elements of risk in the project (construction delay, fuel price escalation, technology performance and so on) must be mitigated or hedged through adequate contracts with builders, fuel suppliers, equipment vendors and others. As a result of the power purchase contract a large portion of the project risk, such as uncertain customer demand, is in fact transferred from the project's equity investors onto the utility (and ultimately the utility's ratepayers). Furthermore, the rigorous risk-management measures regarding construction, fuel risk and so on all come at a cost, which may also be ultimately paid by the purchaser, that is, the utility and its ratepayers. The result of this may be that equity owners do not shoulder a substantial portion of project risk; and in such a case, it might also mean that investor risk (and hence their required rate of return) does not increase proportionally to the level of debt leverage undertaken.

Regardless of the reasons, experience suggests that energy project developers strongly prefer the use of debt over equity as a means of reducing overall project costs. However, wind energy projects in the USA have been characterised not only by a higher cost of equity capital than conventional power plants, as described earlier, but also by a higher proportion of equity in the capital structure. Kahn (1995) cites the example of a 1994 project which was structured using 50 per cent debt and 50 per cent equity. This may be attributed in large part to the presence of the wind-energy production tax credit (PTC) in the USA, which provides a tax credit per kWh of electricity generated by wind power and indirectly encourages increased equity financing.[5] However, even in the absence of the PTC, the proportion of equity in wind energy projects is higher than that in conventional fossil fuel projects.

This need for a higher proportion of equity in wind projects brings us back to the issue of risk. Lenders may not only raise the interest rate required for loans to risky projects, but they will also restrict the amount of credit they are willing to provide. Thus, lenders will determine the level of equity required based on the level of perceived risk in a project. The higher the perceived risk, the higher the proportion of equity capital lenders will require in order to limit the likelihood of default.[6] Standard & Poor's project viability assessment criteria, discussed earlier, also highlight these debt/equity considerations under the 'project structure' risk category. Thus, gas-fired IPP projects in the mid-1990s had typical equity requirements of around 15 per cent, while equity requirements for similar renewable energy projects were often as high as 25 per cent (Brown and Yuen, 1994), leading to much higher financing costs for renewables.

Again, however, such perceptions of risk and associated equity requirements are highly country-dependent. The high proportion of equity required for wind projects, discussed above, is characteristic of the USA and will typically be true for any country not having much experience with wind energy. In countries with strong policy support for wind energy, the cost of capital and required equity fraction can be much lower for two reasons. First, a supportive policy environment reduces investors' perceptions of risk, thus lowering lenders' requirements for a significant equity investment and reducing equity investors' required rate of return. Second, a country's wind energy policy may specifically facilitate the easing of financing conditions through soft loans, for example.

Denmark and Germany, two of the world's leading wind energy producers, highlight each of these two factors. Denmark's support for wind energy comes in two primary forms: a guaranteed purchase price for all wind-generated electricity and an exemption from $CO_2$ and fossil-fuel energy taxes. The market for wind energy in Denmark is secure enough that a new wind turbine installation can be financed through commercial loans with 100 per cent debt at a nominal interest rate of only 5.5–7 per cent per year (Andersen, 1998). Germany's support mechanism for wind power includes an explicit preferential finance component. A German federal funding institution provides loans at below-market interest rates which can cover approximately 75 per cent of project costs. Combined with further commercial loans from local banks, German wind projects' equity contributions are limited to less than 10 per cent (Lindley, 1996). Chapter 7 provides more detailed policy discussions on these and other countries. The critical conclusion, however, is that financing considerations, including the cost of capital and hence the ultimate cost of the project, are intimately linked with the policy environment in place.

### Debt service coverage ratio

The debt service coverage ratio (DSCR) is defined as the ratio between the cash flow arising from a project in any given year and the amount of cash necessary to service all debt payments in that year. The DSCR numerator includes pre-tax income minus operating expenses (O&M, land, insurance, property taxes and so on), while the denominator contains both interest and principal payments. However, in corporate finance, bonds are typically rolled over at the time of maturity such that the principal is never actually paid. Thus in corporate finance, the denominator of the DSCR would usually include only interest payments, while in project finance, the denominator includes both interest and principal payments.

Because the DSCR indicates a project's ability to meet its annual debt service obligations, project-financed projects must pay particular attention to the DSCR. In project finance the lender does not have recourse to the project owner's assets outside of the project, so the lender must ensure that the project itself always has sufficient cash flow in any given year to meet that year's debt payments. Therefore, a lender will never allow an expected DSCR of less than 1.0 in any year. The required minimum DSCR will depend on the project's perceived riskiness and how sensitive the project's cash flow is to uncertain factors such as technology performance, fuel availability and so on. If all project risks have been adequately hedged, then a minimum DSCR of only slightly above 1.0 may be acceptable to the lender. Wiser and Kahn (1996) estimate the minimum acceptable DSCR of an IPP gas-fired project to be around 1.2–1.25, while the minimum DSCR for wind projects is likely to be around 1.4, reflecting the higher real and perceived risks of wind projects. Wind-resource availability risk is particularly important in raising the minimum DSCR; while gas fuel availability can be ensured through an appropriate gas supply contract, wind availability cannot be guaranteed. Some renewable energy projects (not necessarily wind) can have minimum DSCRs of as high as 2.5 (Brown and Yuen, 1994).[7]

The key influence of the minimum DSCR requirement is that it limits the amount of debt which a project can take on, thus raising the necessary amount of (more expensive) equity capital. Due to wind energy's capital-intensive nature, DSCR constraints are most keenly felt in the first years of a project's operation. This can be mitigated if the payment stream received by the wind project is front-loaded, but the utility purchasing the wind power may not necessarily agree to this owing to the increased risk entailed for the utility. One commonly-used front-loading technique involves the use of a constant nominal power purchase price throughout the life of the project. This results in a higher real purchase price in the project's early years, with subsequently lower prices in later years as the constant nominal price loses value through inflation. Another possibility is to back-load project debt repayments such that, rather than a standard mortgage-style repayment scheme of uniform annual payments, debt payments increase over time, thus reducing the debt-service burden in the project's early years.

Debt maturity has a similar impact on finance costs due to its effect on the DSCR. Short-term debt requires higher annual debt payments, resulting in further DSCR constraints. As a result, as the loan term is shortened, a project must rely on a higher proportion of costly equity capital to meet the minimum required DSCR, thus raising project costs. This is mitigated to some degree by the fact that shorter-term loans typically charge a lower interest rate. Nevertheless, overall finance costs increase as loan maturity decreases, due to the DSCR constraint.

The authors used Wiser and Kahn's (1996) project finance pro forma cash-flow model with slight modifications to reflect Danish power purchase conditions and obtained the following results. Assuming the typical Danish wind power purchase price (including energy tax and $CO_2$ tax refunds) of 0.60 DKK/kWh (0.091 US$/kWh at the average 1997 exchange rate) and 100 per cent debt financing (assuming 50 per cent 5-year loan at 6% and 50 per cent 10-year loan at average 7%), it would yield a minimum debt-service coverage ratio of 1.46,[8] which is well in line with typical DSCRs for US wind projects. This is in fact a fictional comparison, because Danish loans are structured differently from US loans; the entire loan is typically repaid before the wind project owner retains any profits. Nevertheless, the comparison illustrates that the high power purchase price of the Danish wind power market appears to allow Danish wind projects to meet typical American debt-service coverage requirements without any need for owner equity input. Wind power projects are therefore highly profitable for owners in Denmark, largely explaining their popularity as investments.

**Overall wind power financing constraints**

The previous discussion highlighted two primary constraints commonly found in financing wind power projects: the high cost of equity capital and the need for a large proportion of equity rather than debt capital. The need for a high equity fraction is in turn influenced by factors such as the debt-service coverage ratio and debt maturity terms. In addition, wind energy projects in the USA pay a higher interest rate on debt than do comparable fossil-fuelled projects. Wiser and Kahn (1996) analysed the difference between typical financing conditions for project-financed wind and gas-fired power projects and analysed to what degree the levelised cost of wind energy could be lowered if wind projects had the same financing terms as gas-fired projects. Their results are summarised in Table 5.1.

The results show, for example, that if the cost of wind energy projects' equity capital could be reduced from the 18%/yr level assumed

**Table 5.1** Comparison of typical financing terms for US wind and gas power projects

| *Financing variable* | *Wind power project* | *Gas-fired project* | *% Reduction in levelised wind energy cost if wind financing terms were equal to those for gas-fired projects* |
|---|---|---|---|
| Debt interest rate | 9.5% | 8.0% | 4 |
| Debt maturity | 12 years | 15 years | 5 |
| Minimum DCSRr | 1.4 | 1.25 | 3 |
| Cost of equity capital | 18% | 12% | 18 |
| All of above variables same for wind as for gas | | | 26 |

*Source*: Wiser and Kahn (1996), p. 41, adapted by author.

for US wind projects down to the 12%/yr level typical for US gas-fired power projects, then the levelised cost of wind energy would be reduced by 18 per cent. The Table 5.1 results were obtained by individually modifying each of the four parameters. The cost reductions are not additive because there are interactive effects between parameters. For example, lowering the interest rate, lengthening the debt maturity and reducing the minimum DSCR requirement all allow reduced reliance on equity capital, thus lessening the impact of reducing the cost of equity capital. An optimal combination of all four factors, calculated by the author, would allow a total reduction in wind-power levelised costs by 26 per cent. Thus, given Wiser and Kahn's assumptions, wind-power levelised costs could be reduced from US$ 0.0495/kWh to US$ 0.0365/kWh if wind power plants received financing terms identical to those of gas-fired power plants.

Thus, wind energy costs can be reduced substantially if the level of risk facing investors is reduced. Some factors, such as wind's inherent resource variability, may mean that wind projects' required DSCR may never become as low as that for gas projects (or may require very conservative wind-resource assessments in return for a low DSCR). However, other factors show considerable room for improvement. For example, the high equity cost of wind projects in the USA is partly a reflection of past technology failures and is not necessarily justified by the reliability of today's state-of-the-art turbines. Wind power financing terms have in fact already improved considerably since the 1980s and should continue to improve as wind-power becomes more accepted as a mainstream energy resource.

Nevertheless, the level of risk experienced by investors is, to a significant degree, a function of the policy environment in place. Policy makers thus have the ability to substantially influence the cost of wind energy, not only through subsidies and incentives, but also by simply improving policy stability and enhancing the availability of reliable power purchase contracts. The challenge for policy makers is to reduce the level of investor risk without simultaneously reducing the incentive for further innovation and cost reduction. The discussion in Chapter 7 demonstrates that mechanisms are available to achieve this and are being successfully implemented in several countries.

## Financing considerations in emerging economies

Though the most significant experience in implementing wind energy projects has been in developed countries, interest is increasing in developing countries as well. India already has one of the world's largest wind power programmes, and China, Egypt and others are also installing significant wind capacity. Nevertheless, developing countries encounter added financing challenges due to the greater risks typically involved.

The most important risks facing investors in emerging economies stem from weaknesses in the institutional and legal frameworks present, which can manifest themselves as political instability, lack of regulatory transparency, corruption, forced contract renegotiation, currency devaluation, labour unrest and many other forms. One risk of particular concern is often the power purchaser's credit strength. In many countries, utilities are an arm of the government and subject to political control, whether in the form of subsidised tariffs, mandates for unprofitable rural electrification or labour rigidity. As a result, the utilities which sign power purchase contracts with power producers may themselves not be financially sound, thus endangering the reliability of the long-term revenue stream which forms the basis of the project. Revenues may also be at risk due to technology performance risk, whether as a result of construction shortcomings or of poor maintenance. Even when profits are realised, governments may sometimes restrict developers' ability to repatriate profits out of the country.

The presence of such risks raises the cost of capital and hence project costs. And though many risks can be reduced through appropriate contractual arrangements and loan guarantees, these too come at a cost. Adding to the problem of high costs is that financing may at times not be forthcoming at any price. In particular, developed countries' commercial banks, having been badly burnt by the Latin American debt crisis of the 1980s, have significantly reduced their exposure to long-term developing country debt, precisely of the type required for power plant development. And though foreign equity capital has become more available in the 1990s, this is also subject to fashion; and the Asian financial crisis of 1997 may reduce foreign investors' enthusiasm for emerging economies for some time to come.

None of the above-mentioned risks are in any way unique to wind power or renewable energy in general. Conventional power plants are equally confronted with the same risk factors. However, wind and other renewable energy projects are at a particular disadvantage because of their relatively small size compared to conventional projects. All power development projects involve substantial transaction costs in terms of contract negotiations and risk management provisions; and these costs are typically similar whether a plant has a 100 MW or 1000 MW capacity. Significant economies of scale therefore exist regarding transaction costs, and wind power plants are typically at a disadvantage. This transaction cost disadvantage for renewables exists in developed countries as well but is particularly severe in developing countries because of the more stringent risk-management needs.

Though financing projects in emerging economies does entail added complexities, a significant number of organisations exist to provide assistance. There are five primary multilateral development banks working in emerging economies: the World Bank, Inter-American Development Bank (IDB), Asian Development Bank (ADB), African Development Bank (AfDB), and European Bank for Reconstruction and Development (EBRD), all of which provide extensive finance for energy projects, of which renewable energy is one component. With the exception of EBRD, these multilateral banks provide loans primarily to governments and require sovereign guarantees of repayment. However, the banks also have separate affiliates which provide finance to the private sector without government guarantees, including the International Finance Corporation (IFC, part of the World Bank Group) and the Inter-American Investment Corporation and Multilateral Investment Fund (both affiliated with the IDB). Other multilateral banks include, among others, the European Investment Bank, Nordic Investment Bank and Islamic Development Bank.

The multilateral development banks are important not only for the financing which they directly provide, but also for ensuring project soundness and thus stimulating further financing from other institutions, such as commercial banks. The World Bank also provides loan guarantees against *force majeure* events to private lenders in order to encourage further private-sector loans. In addition, the Multilateral Investment Guarantee Agency (also part of the World Bank Group) provides investment guarantees regarding currency transfer, expropriation, war, civil disturbance and breach of contract by the host government (Razavi, 1996).

The above-mentioned institutions do not necessarily emphasise renewable energy in their energy-sector loan portfolios, but renewable energy has generally begun to receive a higher profile in recent years. Funds which specifically target renewables include the IFC's Renewable Energy and Energy Efficiency Fund, and the Global Environment Facility (jointly run by the World Bank, UN Development Programme and UN Environment Programme). The European Union also has various programmes including the European Development Fund (Directorate General VIII), EC Investment Partners (DG I), Joule/Thermie (DG XII and XVII), and Altener, Synergy and INCO (DG XVII) programmes (Windpower Monthly, 1998a).

In addition, as part of the Kyoto Protocol to the United Nations Framework Convention on Climate Change, a Clean Development Mechanism is being established through which developed countries (who are obligated to reduce their emissions) can finance climate change mitigation projects such as wind energy in developing countries (who are not obliged to reduce emissions). By participating voluntarily, the developing country receives funding to pursue a less polluting development path; and in return, the developed country receives credit for the emission reductions to contribute towards meeting its own emission reduction commitments (UN, 1997).

Private-sector specialised investment funds also exist for financing energy projects in emerging economies, including the Global Power Investment Company, Scudder Latin America Trust for Independent Power, The Asian Infrastructure Fund, AIG Asian Infrastructure Fund and Alliance ScanEast Fund (Anayiotos, 1994), but such investment funds mostly favour conventional energy projects over renewables. Other examples of funders for renewable energy projects include the German Investment & Development Company, E & Co., Energy Capital Holding Company International, Energy Investors Funds Group, Environmental Enterprises Assistance Fund, Netherlands Development Finance Company, and Impax Capital (REPSource, 1998).

A large number of countries also provide bilateral development assistance for energy projects, many of which include a significant component for environmental protection, including renewable energy. Japan is the largest overall donor in absolute terms, and its development agencies include the Overseas Economic Cooperation Fund, the Export-Import Bank of Japan, and the Japan International Cooperation Agency. The USA is the second-largest donor in absolute terms, and its agencies include the US Agency for International Development, the US Export-Import Bank and the Overseas Private Investment Corporation. Germany's large development assistance programme is implemented by, among others, the Bundesministerium für Wirtschaftliche Zusammenarbeit und Entiwicklung (BMZ), Kreditanstalt für Wiederaufbau (KfW), and Gesellschaft für Technische Zusammenarbeit (GTZ). Other countries with development assistance programmes typically emphasising environmental protection include Canada, Denmark, Finland, Norway and Sweden. Most other European countries' bilateral development programmes include an energy component as well. Many Arab states in the Persian Gulf area also provide significant development assistance for energy projects. Good detailed listings of multilateral, regional and bilateral funding organisations can be found in Razavi (1996) and Private Power Executive (1997).

Bilateral programmes for wind energy provide assistance not only for power plants but also for developing a general wind energy infrastructure in areas such as wind resource assessment, turbine testing, training and so on. An example of such technology transfer co-operation is the Hurghada Wind Energy Technology Center in Egypt, developed through an Egyptian–Danish co-operation programme on renewable energy and providing an important component in Egypt's plan to install 600 MW of wind turbine capacity by the year 2005 (Hansen et al., 1997).

## Competition and power markets

If there is one single development in the electricity industry which most exemplifies the decade of the 1990s, it is probably the increasing trend towards privatisation and competition in power markets. Electric utilities throughout the world have traditionally been viewed as natural monopolies and have operated as either state-owned and -run entities (for example, France, the UK) or as private monopolies operating under close government regulation (for example, the USA, Japan). Developing countries' utilities have been mostly state-owned.

This began to change starting around 1978 and the passage of the US Public Utility Regulatory Policies Act (PURPA), which mandated that US utilities purchase electricity generated by qualifying private power producers. PURPA thus broke US utilities' local monopolies on electricity generation, while maintaining their monopoly in transmission, distribution and sales. Since that time, around the world, utilities' hold on the generation market has steadily weakened as independent power producers (IPPs) rapidly expanded their reach. Nonetheless, in spite of private power's encroachment, this model of competition has been characterised by the utility continuing to maintain responsibility for overall system planning, power plant dispatch and system reliability. Thus, the utility remained the sole outlet for all generators' power output. This model of competing generators selling to a common purchasing agent has been termed the 'Purchasing Agency' model (Hunt and Shuttleworth, 1996). This Purchasing Agency model turned out to be highly beneficial for wind energy, as it was under such systems that wind energy first began to thrive, either through long-term power purchase contracts with utilities (as in the USA) or through implicit contracts backed by government mandates (as in Denmark).

Beginning in the early 1990s with Norway, the UK, Chile, Argentina and others, countries began pushing competition further by eliminating the common purchasing agent and allowing all generators and power marketers to sell directly to either wholesale or retail customers. In this new model, sellers are granted non-discriminatory access to transmission and distribution systems, with plant dispatch performed by an independent system operator. Contracts between generators and customers are either negotiated bilaterally or accomplished through a power pool. In such wholesale and retail competition systems, the transmission and distribution 'wires' businesses continue to operate as regulated monopolies, but all other services are open to competition, often requiring utility divestiture of generation assets to reduce utility market power. Thus, the utility's role in the generation market is reduced even further than in the earlier Purchasing Agency model; and importantly, the utility no longer fulfils the centralised generation-system planning role. In fact, the generation planning role is eliminated entirely, as it is assumed that the 'market' will adequately anticipate growing power needs and will build new power plants accordingly to meet this need.

For both utility and non-utility electricity generators, this change is nothing short of revolutionary. The replacement of stable rate-of-return regulation with all-out competition means that the days of fixed profits guaranteed by long-term power contracts are rapidly disappearing. What implications does this hold for the development of new power plants, and, more specifically, what is the prognosis for wind power in this new environment?

First and foremost, wholesale and retail competition mean a drastic increase in risk for generators. Increases in risk are inevitably accompanied by increases in the cost of capital, as discussed earlier in this chapter. This in turn means that investors increasingly favour generation technologies with low capital costs, such as gas turbines, sometimes even despite increased operating costs. Wind energy, which is characterised by high capital costs and minimal operating costs, is thus hurt by the shift to low capital-cost technologies. Wind power does have a compensating advantage of being small and modular, however, allowing investments in small increments and thus reducing the magnitude of risks involved.

Nuclear power, being both capital-intensive and large-scale, has perhaps been the greatest casualty of this increased risk aversion.

Few investors are willing to risk the billions of dollars needed to construct a nuclear power plant without guarantees of stable revenues; and in fact even already-existing nuclear power plants in places like the UK and California have remained financially viable in their new competitive markets only as a result of price supports and special 'transitional' financing provisions. Large-scale hydroelectric power has had similar difficulties attracting financing for new projects because of its highly capital-intensive nature. Argentina, for example, which previously derived the bulk of its electricity from hydro power, has witnessed a dramatic shift to gas-fired generation after the creation of its competitive market (Hasson et al., 1998).

In spite of its rapidly improving economic viability, wind power is in general not yet able to compete head-on with conventional electricity generation, particularly given the low natural-gas prices which have prevailed during the 1990s. Given the extreme difficulty of financing wind power projects in the absence of a guaranteed long-term revenue stream, the move away from long-term power purchase contracts presents a serious challenge for continued wind power development. This is particularly true for the non-recourse, project-financed, IPP development mode; and wind power development may move increasingly towards a corporate-financed world supported by larger companies' strong balance sheets. However, the overall prognosis for wind power is by no means all bad. The following sections describe a variety of considerations of competitive markets and the opportunities as well as challenges they entail for future wind power development.

## Common purchasing agency vs. wholesale or retail competition

The first issue which requires clarification is regarding the nature of competition in the electricity industry. Power sector reform and the introduction of competition means very different things in different countries. Hunt and Shuttleworth (1996) describe four models of electric industry structure, alluded to earlier: (1) Monopoly, (2) Purchasing Agency, (3) Wholesale Competition, and (4) Retail Competition. The trend in developed countries is towards either Model 3 or Model 4, with generators and marketers competing to sell power directly to either wholesale or retail customers, either through bilateral contracts or through a power pool. In developing countries and emerging economies, however, the picture is typically quite different.

Turkson (2000), for example, carried out a detailed study of power sector reforms in sub-Saharan Africa which revealed that, while many sub-Saharan African countries are in the process of restructuring their electricity markets, the currently envisaged nature of competition is almost exclusively of the Model 2 Purchasing Agency model in which IPPs vie for long-term power purchase contracts with the otherwise monopoly utility. Only in one sub-Saharan country, Ghana, are there current plans to move beyond the Purchasing Agency model to the Wholesale Competition model.

In Asia, the picture is no different. Independent power generation is well-established throughout East, South-east, and South Asia, all of whom underwent a major private power boom in the 1990s. In spite of this emphasis on increased privatisation and competition throughout the region, competition has exclusively entailed the use of the Purchasing Agency model. Wholesale or retail competition are, at most, still on the distant horizon throughout Asia. Of the world's developing regions, only in Latin America has there been significant movement beyond the Purchasing Agency mode, led by Chile and Argentina who have been amongst the world's pioneers of electricity market restructuring.

With the exception of a few countries, therefore, competition throughout the developing world is unlikely to move beyond the Purchasing Agency model within the next decade. India and China, perhaps the two developing countries with the most ambitious wind energy plans, also show no signs of dismantling their existing monopoly structure beyond encouraging IPP development. This Purchasing Agency model, encouraging IPPs to sell power to the otherwise monopoly utility, is precisely the model under which wind power development first began to flourish in developed countries. Because the Purchasing Agency model involves continued planned generation expansion in a centralised, systematic way, interested regulators and governments can readily stimulate wind energy development through the consideration of non-price factors (environmental externalities, fuel diversity and so on) in the system-planning process and through the creation of stable power purchase regulations. Therefore, the advent of competition, independent power development, and the break on utilities' generation monopoly is likely to provide a boon to wind power development in developing countries who follow the Purchasing Agency competition model.

In developed countries, the move to wholesale and retail competition entails further complexities for wind power. With the elimination of the centralised generation-planning process, competition amongst generators tends to move towards one based exclusively on price, in which wind energy cannot readily compete at this time. Perhaps more importantly, any mandates by regulators for utilities to purchase above-market priced wind (or other) energy could force an increase in utility tariffs and hence harm the utilities' competitiveness against other power sellers. Utilities competing for wholesale and retail customers are therefore likely to actively resist power purchase mandates for wind power which they may have previously accepted under the Purchasing Agency competition model. Some policy makers may also have philosophical objections to special treatment for any energy source through mandatory power purchase contracts, arguing that the point of open competition is precisely to eliminate such market distortions.

### Renewables market set-asides

Yet, countries moving to wholesale and retail competition do continue to find reasons to support renewable energy and have developed support mechanisms which are compatible with the competitive market. The most well-known such mechanism is the Non Fossil Fuel Obligation (NFFO) of the UK, which sets aside a certain portion of the electricity market to be filled by renewable energy, including wind (see Chapter 7 for details). The NFFO allocates long-term power purchase contracts to renewable generators based on a competitive bidding process, thus maintaining the discipline of market forces while sheltering renewables from the full brunt of open competition against conventional fuels. Importantly, the NFFO pays renewable generators the premium between their contract price and the open-market price out of a special levy charged to all electricity consumers, thus eliminating any competitive disadvantage for those utilities purchasing the renewable energy.

Similarly, the California electricity market provides special funds to support renewable energy using a competitive framework and based on a non-bypassable 'system benefits charge' levied on all electricity users. Other mechanisms to support renewable energy within a competitive framework include the Netherlands' Green

Labels programme and the similar proposed Renewables Portfolio Standard in the USA, both of which involve use of tradable renewable energy credits and place a requirement upon all retail electricity suppliers to purchase a certain proportion of their electricity from renewable generators (see Chapter 7).

Thus, wholesale and retail competition do not preclude the use of special support mechanisms for renewable energy, including wind. The key to such mechanisms is that they should affect all market players equally so as not to create any unfair competitive advantages. Furthermore, the long-term goal of any such renewables support mechanism in the competitive market must be to eventually phase out the need for such special support and to bring about viable open competition amongst all technologies. Two primary conditions are required for this to happen. First, the market must move towards reflecting the full benefits of renewable energy not currently reflected in the market price, including the value of environmental benefits. Secondly, the support mechanism must encourage steady and sustained cost reduction amongst renewable energy technologies. Issues regarding the environment and competition on the open market are discussed further below.

### Environment

Wind energy provides significant environmental benefits compared to conventional electricity sources, including reduced local air pollution and reduced emissions of gases contributing to global climate change. Wind energy can also entail certain environmental disadvantages such as greater visual intrussion or accidental avian deaths. In most countries, neither these advantages nor disadvantages are reflected in the price of electricity. As discussed in Chapter 6, the balance of evidence strongly suggests that wind energy's environmental benefits far outweigh its damages. Thus, lack of reflection of environmental impacts in electricity prices means that wind energy is undervalued in the power market.

Within the centralised generation-planning mode of utility monopolies and Purchasing Agency-based competition, energy regulators have sometimes tried to address this market failure by providing wind energy and other renewables with certain credits or subjective high 'point scores'. When the generation planning function is eliminated in a fully competitive market, however, such adjustments become inapplicable. In such a case, the failure of market prices to reflect environmental impacts becomes untenable and provides a severe competitive disadvantage to wind energy and all other non-polluting renewable energy technologies. As long as such externalities are not properly reflected in market prices, special support mechanisms for renewable energy will continue to be justified.

The great unknown in the debate over environmental externalities, however, is the issue of global warming. Wind energy's lack of greenhouse gas (GHG) emissions, responsible for global warming, may well represent its greatest environmental advantage over fossil fuels. Yet, the near impossibility of accurately quantifying the damage costs of global climate change means that no accurate and widely acceptable estimates exist of wind energy's environmental benefits. This lack of agreed-upon values is perhaps the greatest obstacle to getting market electricity prices to more fully reflect environmental impacts.

However, the United Nations Framework Convention on Climate Change and its subsequent 1997 Kyoto Protocol do provide an indication of a way forward. The Kyoto Protocol commits industrialised 'Annex One' countries, for the first time ever, to binding reductions in greenhouse gas emissions (UN, 1997). Though the Protocol is itself highly contentious and its eventual entry into force still in doubt as of early 1999, the political climate-change negotiation process is nevertheless moving inexorably in the direction of binding emission reductions. Once such binding reductions come into effect, this will automatically create a defacto market reflecting participants' willingness to pay for projects which achieve GHG emission reductions, and hence placing a monetary value on $CO_2$ abatement. Thus, even in the absence of broadly accepted climate change externality values, wind energy projects could well begin receiving actual payments in some form for GHG reductions within the next five to ten years. Such GHG reduction payments could potentially provide a significant boost to wind energy's financial viability.

## Competing on the open market

Looking beyond the next decade or so, separate markets specially reserved for renewables, such as the UK's NFFO, may not be politically sustainable over the long run. Eventually, wind and all other forms of renewable energy will probably have to compete against conventional energy technologies in an open, deregulated generation market. Thus, continued cost reductions will be necessary for any renewable energy technology to survive over the long term. Chapter 4 demonstrates that wind energy is indeed becoming increasingly economically viable, even in the absence of consideration of environmental benefits. Yet, what is the prognosis for wind energy truly competing against conventional technologies?

Some of the most promising news to date in this regard was highlighted in Chapter 4 regarding the 1998 integrated resource-planning process of Northern States Power in Minnesota, USA, where wind energy was found to be the least-cost generation option, with expected lifetime costs potentially even lower than those for combined cycle natural gas generation.

Other promising news comes from Ireland and the results of its Alternative Energy Requirement (AER III) bidding process. The lowest bid price received for a wind energy project was 2.21 Irish pence per kWh, or approximately 0.028 ECU/kWh or 0.031 US$/kWh, for a 15-year power purchase contract. The contract would commence within 1999, and the power purchase price would escalate over time in line with a price index (O'Gallachoir, 1998). The project would receive a grant of ECU 80 000 per MW installed (approximately 90 US$/kW) and possibly some Irish tax credits as well; and even then some observers believe this bid price to be too low to be viable (Windpower Monthly, 1998b). Nevertheless, a bid price of 0.031 US$/kWh is perhaps the lowest price for wind energy yet seen anywhere in the world and reflects a price level fully competitive with that of coal, gas and nuclear electricity (see Chapter 4).

While the low price of the Irish AER III does reflect the ECU 80 000 per MW capital subsidy, this subsidy is fairly modest, representing less than one-tenth of the project's total capital cost. Furthermore, it must be kept in mind that all energy industries (coal, gas, nuclear and so on) have been and continue to be major recipients of government subsidies. In the USA, for example, the US Energy Information Administration calculated that the coal, natural gas and nuclear energy industries each received subsidies of between US$900 million and 1.15 billion in 1992, while renewable energy received somewhat less (EIA, 1992).

While the Irish AER III results do not necessarily indicate wind energy being fully competitive with conventional power, they do nonetheless provide further evidence that the day of wind energy successfully competing head-on in the open market is approaching and could happen within the next ten years in areas with favourable wind conditions. On the other hand, overly rapid price reductions are not necessarily a good sign for the long-term health of the wind industry. Wind turbine manufacturers must be able to earn profits if they are to continue to invest in developing new technology. The low Irish bid prices were therefore greeted with decidedly mixed reactions by the wind industry.

It should also be noted that the low Irish price was achieved through a bidding process. While bidding schemes like the UK's NFFO have had their share of difficulties (see Chapter 7) and have not always been cheap due to high up-front transaction costs, they have been generally successful at stimulating significant cost reductions. Continued cost improvements will be necessary to sustain the future growth of the wind energy industry, and support schemes need to provide appropriate incentives for this.

### Forward markets, spot markets and bilateral contracts

Overall costs are only part of the picture, however. Electricity generated during times of peak demand is much more valuable than that generated during times of low demand. In this regard, wind energy faces two drawbacks compared to technologies such as gas turbines. First, wind energy is not dispatchable; it is available only when the wind is blowing. Thus, payments received by wind energy generators may be low if wind availability does not coincide with times of high demand. Secondly, and perhaps more importantly, the lack of dispatchability could complicate wind generators' ability to function in the forward and spot markets characteristic of competitive generation systems.

The question addressed in this section is therefore as follows. *Assuming* that wind energy becomes fully competitive on a total-cost basis with conventional technologies, will wind energy's intermittent nature hamper its viability in the generation markets which prevail under wholesale and retail competition?

To address this, it is necessary first to provide some background on the functioning of competitive generation markets. With wholesale and retail competition, electricity generators can sell their output through two broad types of sales: bilateral contracts, and short-term forward and spot power sales. Bilateral contracts are negotiated directly between a buyer and seller and can be of any duration: weeks, months or years. They are normally of a longer-term nature and provide stability to both the seller and purchaser in terms of quantity and price. Forward and spot markets are short-term markets which typically operate on an auction system similar to a stock exchange. Forward markets normally operate on a week-ahead, day-ahead or hour-ahead basis, while spot markets provide instantaneous matching of supply and demand. The short-term markets serve two critical functions. First, they establish a transparent 'market' price for electricity at any given time. Secondly, the spot market compensates for any imbalances between the level of sales contracted for in bilateral and forward contracts and the level of electricity actually consumed.

Some confusion appears to exist between the definition of forward markets and spot markets. Some observers choose to characterise short-term forward markets (day-ahead and hour-ahead) as spot markets. In such a case, the 'spot market price' would typically refer to the day-ahead market clearing price. For the sake of this discussion, we generally classify day-ahead and hour-ahead markets as forward markets and refer to spot markets primarily in terms of instantaneous matching of supply and demand at the time of delivery. Nevertheless, the terms are used somewhat interchangeably, and we refer to the short-term forward market and the spot market collectively as 'short-term markets'.

Bilateral contracts are negotiated directly between the generator and a purchaser (an electricity wholesaler or retailer). This purchaser would normally purchase electricity from a variety of generators who have different operational and cost characteristics. A purchaser's contract with a wind energy generator would therefore be only one of many contracts signed with different generation sources. Because wind is an intermittent resource which cannot be accurately predicted far in advance, a long-term contract with a wind generator would not be useful for the purchaser to lock in any firm quantity of electricity at any particular time. If the wind is not blowing at the given time, the purchaser would have to make up for the lack of wind electricity by either purchasing power from another dispatchable generator (such as hydro or a gas turbine) through another bilateral contract, or more likely by purchasing from the short-term forward or spot market. This need to rely on the short-term markets makes wind power relatively unattractive for long-term contracts. The purchaser would be likely to demand a power purchase price well below the average spot market price at the time of wind availability. The wind generator, on the other hand, should be able to sell its power on the spot market for around the spot market price and would therefore have little incentive to sign such a low-priced contract.

This does not mean that wind power is entirely unsuited to long-term bilateral contracts, however. First, a wind generator could team with a dispatchable higher-cost generator to provide firm power at all times. The dispatchable generator could be used to make up for any shortfall in output from the wind generator.

Secondly and more importantly, however, purchasers may be interested in signing long-term contracts with wind generators not so much to lock in a particular quantity of energy at a particular price, but rather to lock in $CO_2$ emission reduction credits at a particular price. As mentioned earlier, the UN Framework Convention on Climate Change process is moving towards mandatory $CO_2$ emission reductions, which will inevitably lead to a market for $CO_2$ credits. In fact, the Green Labels programme in the Netherlands, which requires all Dutch utilities to purchase a certain quantity of green 'labels' generated through renewable electricity (see Chapter 7), is already such a market. Under this programme, Dutch utilities are signing contracts with renewable generators to lock in what are essentially $CO_2$ emission reduction credits. As more countries begin implementing mandatory $CO_2$ reduction programmes, this $CO_2$ credit market could well become the leading force for stimulating long-term bilateral energy contracts for wind generators.

Short-term forward and spot markets are organised through a power exchange which matches bids to buy and sell power. Detailed operation differs between countries, but the basic operation can be described as follows. The forward and spot markets and the power pool through which they operate provide two related but distinct functions: (1) determination of which power plants to dispatch at any given time, and (2) determination of the market price for short-term power sales and purchases at any given time. These two functions may be carried out by one or by separate entities. In the

England and Wales Power Pool, for example, dispatch decisions as well as establishment of pool prices and settlement of contract imbalances (between kWh contracted for delivery and kWh actually consumed) are all carried out by the National Grid Company (NGC). In California, plant scheduling and dispatch are carried out by the California Independent System Operator, while the short-term trading function and establishment of the market clearing price are handled by the California Power Exchange. In any event, the operations of the dispatching entity and the power exchange entity are closely co-ordinated and, for illustrative purposes, are lumped together as functions of the 'market operator' (Hunt and Shuttleworth, 1996) in this discussion.

The market operator operates week-ahead, day-ahead or hour-ahead forward markets in which generators submit bids a week, day or hour in advance, offering to provide a given quantity of generation at a given price during a particular hour. The market operator ranks the bids for each hour in ascending order of price. Based on the estimated total electricity demand in each hour, the market operator determines which power plants to dispatch in each hour. The bid price of the most expensive generator to be dispatched becomes the market clearing price or pool price paid to all generators dispatched in that hour. Hunt and Shuttleworth (1996) provide a useful narrative description of the functioning of the England and Wales Electricity Pool:

1 A day in advance of trading, generators submit data on the forecast availability of generating sets ('gensets') and the offer price at which they are prepared to generate. The National Grid Company (NGC) prepares a detailed demand forecast for each half-hour of the coming day.

2 A computer program is used to produce an 'Unconstrained Schedule', or 'U-Schedule'. This is a plan of generation which meets forecast demand at least cost (in terms of offer prices), ignoring any transmission constraints.

3 For any half-hour, the offer price of the marginal (most expensive) genset operating in the U-Schedule determines the 'System Marginal Price' (SMP). The Pool Purchase Price (PPP) is

equal to the SMP augmented by an element related to the expected degree of capacity surplus on the system.

4. Any genset capacity offered but not needed in the U-Schedule is awarded an availability bonus, which is also related to the expected degree of capacity surplus on the system.
5. On the day, NGC issues instructions to gensets as to when and how much to generate.
6. Where NGC instructs a generator to deviate from the level of U-Schedule output, the change in output is bought or sold by the pool at each genset's own offer price. Failure to meet instructions, or to be available as declared the previous day, is penalised.
7. After all of these transactions have been completed, the price to consumers (Pool Selling Price or PSP) is calculated as the sum of net payments to generators divided by the total amount actually generated.

*Source*: Hunt and Shuttleworth (1996), *Competition and Choice in Electricity*, pp. 168–9. (Copyright John Wiley & Sons, reproduced with permission.)

Figure 5.1 provides a similar but simpler graphical illustration of the operation of the Nord Pool Nordic short-term market. In the Nord Pool market, electricity suppliers and purchasers each supply bids for the quantity and price of electricity they would like to sell or buy at a particular time. The bids are arranged in order of price to create supply and demand curves. The demand curve can be seen to be very inelastic, with demand increasing only slightly as prices fall. The supply curve, on the other hand, is composed of many different power plants demanding a wide range of prices. Those power plants with the lowest short-run operating costs (hydro and wind, whose operating costs are essentially zero) compose the lowest bids, while the higher bids of the CHP (combined heat and power) and condensing fossil-fuel plants reflect their higher short-run (primarily fuel) costs. The market clearing price is set at the point of intersection between the supply and demand curves, or approximately 19 Norwegian øre per kWh (around 0.025 US$/kWh) in the case of Figure 5.1.

**Figure 5.1** Supply and demand curves for the Nordic electricity system

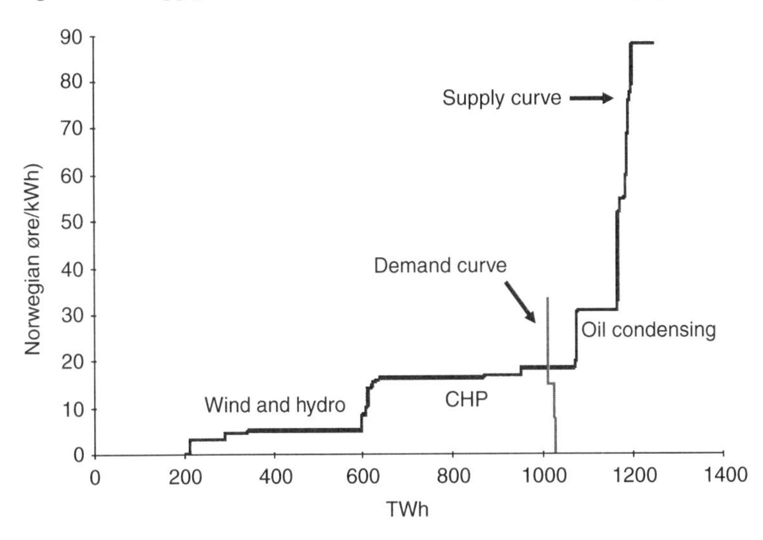


*Source:* Nielsen and Morthorst (1998).

How does wind power perform in such markets? This depends to a large degree on the accuracy with which wind availability can be predicted on a day-ahead or hour-ahead basis. Because power must be bid into the short-term market in advance, accurate prediction of wind plants' output is essential for submitting viable bids.

Wind prediction techniques have improved significantly (see Chapter 3) and are now capable of forecasting wind power output up to 36 hours in advance with an accuracy of around +/– 20 per cent. In other words, if the day-ahead predicted wind power output is, say, 100 MWh in a particular hour, the actual power generated during that hour will reliably be between 80 and 120 MWh. This level of accuracy should generally be sufficient to allow wind generators to bid power into the day-ahead market. Nonetheless, this variability is significantly greater than the typical day-ahead output variability of a fossil fuel-based plant. As a result, the viability of wind power in day-ahead markets is highly dependent on the degree of penalties charged by the market operator to generators who are unable to meet their bid-in electricity commitments. Thus, even when wind power becomes fully competitive on an overall cost basis, wind power's viability within short-term forward and spot markets will be dependent on the particular market rules in place. If generators are charged a significant penalty for delivering less electricity than they promise, or if they are paid a very low rate for surplus electricity delivered in excess of their bid, then wind power will be harmed.

The US natural gas market, for example, does charge heavy penalties to those suppliers who commit to deliveries which they are subsequently unable to deliver. This works well in the gas market because its relatively slow-moving nature allows gas traders to balance their supply and demand bids amongst themselves and thus avoid imbalances within the spot market. However, Hunt and Shuttleworth (1996) argue that the instantaneous nature of electricity markets makes such arrangements impractical for electricity. Stiff penalties, they claim, reduce economic efficiency by stimulating excessive bid adjustments by traders and by discouraging generators from allowing any flexibility in their dispatch, thus making the dispatcher's job exceedingly difficult. Imbalances in electricity markets between bid quantities and actually delivered quantities, they argue, are best handled by market structures specifically designed to handle such inevitable imbalances.

The Nord Pool power market in Scandinavia, for example, eschews penalties and relies on a market-based balancing mechanism. In addition to short-term forward markets, the Nord Pool system operates a market for *regulerkraft* or 'regulation power' in Norway. This regulation market is a market for handling imbalances between the quantities of generation bid, the quantities of consumption bid, and the actual realised generation and consumption. If a purchaser consumes more electricity than it has contracted for, its extra consumption is provided for by this regulation market. Similarly, if a generator delivers more or less electricity than its bid quantity, this surplus or deficit is also absorbed by the regulation market.

The cost of regulation power is related to the pool or spot price and the amount of regulation power required. The price of up-regulation power (to request other generators to produce more to make up for one's own under-production) is normally greater than the spot price itself, while the price of down-regulation power (selling one's excess power on the spot market while requesting other generators to back down) is less than the spot price. In other words, a generator bidding into the forward market will lose money by either over-generating or under-generating compared to its bid quantity. The generator thus faces incentives to generate exactly as much electricity as it has bid.

The operation of the regulation market can best be illustrated through an example. Suppose the pool price has been established at $20/MWh for a particular hour, while the price of up-regulation has been established at $25/MWh; and the price of down-regulation is $17/MWh. A wind energy generator has bid to supply 10 MWh during that particular hour. If the wind generator produces exactly 10 MWh as it has bid, then it would receive $20/MWh · 10 MWh = $200.

Suppose, however, that lower-than-predicted wind speeds mean that the wind generator is only able to deliver 9 MWh. It must then purchase 1 MWh at the $25/MWh up-regulation price to meet its 10 MWh commitment. Thus, the wind generator receives $20/MWh · 10 MWh = $200 from the pool but pays out $25/MWh · 1 MWh = $25, resulting in a net receipt of $175 for 9 MWh of power, or $19.44/MWh. Had the wind generator known from the start that it would only be able to produce 9 MWh and bid accordingly, it would have received $20/MWh · 9 MWh = $180. Thus, the generator loses $5 for having over-bid its delivery by 1 MWh. Note, however, that the pool still pays only $20/MWh · 10 MWh = $200 for the 10 MWh of power. The extra $5 cost of having to resort to up-regulation power is absorbed entirely by the wind generator.

Similarly, suppose that higher-than-predicted wind speeds mean that the wind generator produces 11 MWh instead of 10. For the first 10 MWh, the wind generator receives $20/MWh · 10 MWh = $200. Regarding the extra 1 MWh, the down-regulation market's designated generator receives $20 from the pool for this MWh but does not generate. Instead, the down-regulation generator meets its own generation commitment to the pool by purchasing the wind generator's extra 1 MWh for the down-regulation price of $17. Thus, the wind generator receives a total of $217 for 11 MWh of production, or $19.72/MWh. Had the wind generator correctly bid to produce 11 MWh in the first place, it would have received $220. The wind generator thus loses $3 by under-bidding by 1 MWh. This $3 goes to the generator in the down-regulation market, for whom the $3 represents pure profit. Again, the pool itself continues to pay $20/MWh and is unaffected by the transaction in the regulation market.

**Figure 5.2** Illustrative price of regulating power on the Nord Pool market

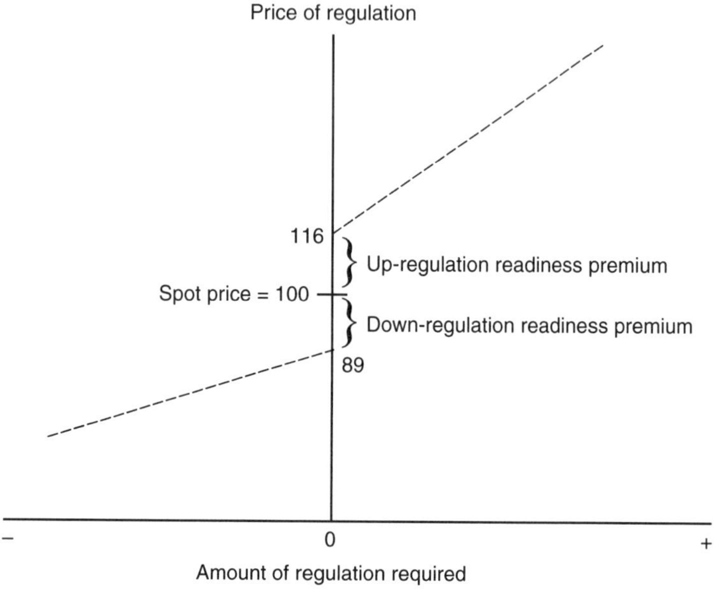

Price of regulation

116

Up-regulation readiness premium

Spot price = 100

Down-regulation readiness premium

89

– 0 +

Amount of regulation required

*Source*: Skytte (1999).

Skytte (1999) analysed the cost of up-regulation and down-regulation in the Nord Pool market. As represented graphically in Figure 5.2, Skytte found that the price of regulating power is a factor of the spot price, a readiness premium, and the amount of regulation required.

Given a hypothetical spot price of 100, the readiness premium for up-regulation might raise the minimum price for up-regulating power to 116, for example. This readiness premium essentially represents the option value of a unit of reserve capacity. The price of up-regulation then increases in direct relation to the amount of up-regulation required. Thus, given the Figure 5.2 spot-market price of 100, purchasing 1 kWh of up-regulating power might cost only 116 per kWh, but purchasing 50 MWh of up-regulating power might cost significantly more per kWh. Similarly for down-regulation, the readiness premium would set a maximum price for down-regulating power, and this price would decrease in direct proportion to the amount of down-regulating power required.

Skytte identified several other characteristics of the Nord Pool system's regulating power market. First, the magnitude of the readiness premium for up- and down-regulation depends on the magnitude of the spot market price. When the spot price is low, the readiness premium for up-regulation is higher than the readiness premium for down-regulation. However, when the spot price is high, the opposite holds true; the readiness premium for down-regulation is higher than the premium for up-regulation. Secondly, Skytte found that the up-regulation price curve has a steeper slope than the down-regulation price curve. In other words, the total amount of regulating power required has a greater influence on the up-regulating price than on the down-regulating price.[9]

These results are based on the existing Nord Pool market configuration and are not necessarily transferable to other markets. Nevertheless, the results indicate that wind generators can develop an optimum bidding strategy based on the level of the anticipated spot price, the variability of wind conditions, and the magnitude of intermittent resources in the market. If wind variability is high or wind prediction accuracy is poor, the amount of required regulation power is likely to be high, which in turn is likely to have a greater impact on the price of up-regulation than on down-regulation in the Nord Pool. Thus, with high wind-resource variability, it is more advantageous for wind generators to under-bid their kWh commitment in the spot market and hence increase their likelihood of requiring down-regulation rather than up-regulation. The same holds true when the time span between the submission of bids and actual delivery increases, as well as when the total amount of wind generation on the market increases. It should be possible to develop similar types of optimum bidding strategies for other non-Nordic markets as well.

Most importantly, the study shows that wind generators in the Nord Pool market can manage their resource variability risks through use of the regulation market and a careful bidding strategy. Nielsen and Morthorst (1998) found that wind power's average marginal cost of regulation power at the time of regulation is approximately 0.03–0.04 DKK/kWh (approximately 0.0045–0.006 US$/kWh). The overall cost of regulation power averaged over a wind generator's total production is approximately 0.01–0.02 DKK/kWh (approximately 0.0015–0.003 US$/kWh) using today's most advanced wind prediction techniques. While these are not insignificant sums, nor are they crippling, representing well under 10 per cent wind energy's generation cost.[10]

In conclusion, the improvement in 24–36 hour-ahead wind prediction techniques, combined with the emergence of short-term forward and spot markets, provide a significant boost to wind energy's competitive potential in the generation market. If wind energy becomes cost-competitive with conventional power on a total cost basis, then its intermittent nature should not cause serious problems for its viability in the short-term competitive generation market. However, wind power's intermittence does make it less conducive to long-term bilateral contracts, making it more difficult for wind generators to lock in a long-term fixed price for its power. This heavy reliance by wind power on unpredictable short-term markets in lieu of more stable long-term contracts could raise financing challenges. On the other hand, an emerging market for $CO_2$ emission reduction credits could provide the necessary impetus for long-term bilateral contracts as well.

Furthermore, the above conclusion about wind power's viability in short-term markets assumes that market operation rules do not impose heavy penalties for imbalances between bid quantities and actually delivered quantities. Markets which do impose severe penalties would greatly impair wind and other intermittent resources' ability to compete. However, as discussed earlier, the instantaneous nature of electricity systems makes severe penalties unconducive to overall economic efficiency for reasons unrelated to wind energy. The overall efficiency of the system, as well as the viability of wind energy, is enhanced by using a market-based system for addressing imbalances, rather than a penalty system.

Other developments also help wind energy's ability to operate in the short-term generation markets. First, electricity markets are becoming increasingly integrated around the world, whether in Europe through the EU liberalisation agenda, in Africa through the Southern African Development Community, or in South America's Mercosur market. These larger markets, with their greater resource diversity and significant presence of highly flexible hydro power, further increase the capacity of the power markets to absorb fluctuating resources like wind. Secondly, the continuing development of options markets and other risk management techniques means that electricity markets are becoming ever more flexible. With this increased flexibility, the markets' ability to handle intermittent resources should again improve. The issues of dispatchability and resource variability are thus likely to decline in importance as competitive generation markets further develop.

### Transmission issues

The discussion on wind energy's prospects in competitive markets has so far ignored the issue of transmission. However, for competing generators, access and pricing for transmission services is a crucial issue. This section provides a brief discussion of transmission pricing in competitive generation markets and its effect on wind energy.

In a competitive market, generators should be charged for transmission services based directly on the actual demands which they place on the transmission system. The impact of such transmission pricing on wind energy depends on the specific type of wind power facility in question. Wind energy advocates often tout the inherent benefits of its dispersed nature which may allow wind turbines to be placed near local load centres. Such benefits are particularly significant in remote communities poorly served by high-voltage transmission networks, where local distributed resources like wind energy could potentially eliminate the need for expensive transmission facilities altogether. Even in less remote fully grid-connected communities, dispersed wind turbines may still have advantages over more centralised conventional generators by feeding directly into the local distribution network and thus bypassing the long-distance transmission network. Thus, in a country like Denmark where wind turbines are scattered throughout the country, these turbines may incur much lower transmission costs when serving local loads than, say, importing hydro power from the distant mountains of Norway.

It is generally more common, however, to install many wind turbines together in concentrated locations, or wind farms, rather than in widely dispersed individual sites. Wind farms typically offer several advantages compared to dispersed wind development, including economies of scale and the ability to make maximum use of wind resources concentrated in specific areas like mountain passes. With such farms, however, wind energy begins to take on the characteristics of larger-scale centralised generation sources which are located away from the load centres and require transmission facilities. In fact, the areas of greatest wind availability are often far away from population centres; and in this sense large wind farms may often most resemble large hydroelectric facilities, located in remote areas and needing even more extensive transmission than fossil-fuelled or nuclear power plants. Many of the best wind resources in the USA and China, for example, are located in these countries' sparsely populated western regions, while many of Chile and Argentina's wind resources are in their similarly remote southern zones.

Because transmission costs typically increase with distance, remote wind facilities are at a disadvantage compared to fossil fuel plants which have more locational flexibility. However, the impact of distance on transmission costs depends to some degree on the type of transmission pricing system in place. With some pricing schemes like 'contract path' and 'megawatt-mile' pricing, transmission charges increase directly in-line with the distance covered in the transmission system. Contract-path pricing is based on the hypothetical transmission distance covered between the generator and its contracted customer, while megawatt-mile pricing uses load flow analysis to provide a more realistic determination of the actual transmission distance likely to be covered given the network configuration and loads on the system. In either case, however, a wind generator located far from load centres is likely to face correspondingly higher transmission charges than a competing more conveniently-located fossil fuel-based plant.

'Postage-stamp' pricing has been the most common transmission pricing scheme in the USA and is the most simple, charging a uniform fee per MW for use of the transmission system within a given zone, regardless of the distance required within that zone. Thus, a 100 MW generator located 1 kilometre from the load, and another 100 MW generator located 100 kilometres from the load would both pay an identical transmission fee as long as both generators were within the same transmission zone. However, if several zones must be crossed between the generator and the load centre, then the postage-stamp price must be paid for the beginning zone, the end zone and all zones in between. Thus with postage-stamp systems, up to a certain distance, transmission prices are unrelated to distance; but as distances increase and begin to cover more than one zone, prices begin increasing in a 'lumpy' fashion in line with distance.

Distance is not the only factor affecting transmission prices, however. Perhaps even more important, in terms of its impact on wind, is the method for pricing firm vs. non-firm transmission capacity. A firm transmission contract provides a generator with guaranteed access to the transmission system regardless of congestion. Non-firm contracts only provide space on the network on an as-available basis; so in the event of transmission congestion, a generator with a non-firm contract may be curtailed and thus unable to sell its power. Because transmission congestion is often (but not necessarily) correlated with times of peak demand and correspondingly high power prices, a generator who is curtailed at such times may risk losing substantial revenue. Firm transmission contracts must typically be signed far in advance, while non-firm contracts are available at shorter notice.

The problem for intermittent generators like wind is that firm transmission contract charges are normally structured per MW of maximum capacity reserved, on a 'take-or-pay' basis. In other words, the generator must always pay for the full amount of capacity it reserves on the transmission system regardless of how much energy it actually transmits. Thus, if a 100 MW generator wants to be sure of being able to transmit its full 100 MW of output at any given time, it would have to buy a firm transmission contract for 100 MW; and the generator would lose money any time it transmits less than the full 100 MW. As a result, intermittent resources like wind with low capacity factors are particularly disadvantaged by capacity-based take-or-pay contracts. To be guaranteed access to the transmission grid at all times, a 100 MW wind facility with a 20 per cent capacity factor would have to purchase 100 MW of firm transmission even if it only generates 20 MW on average.

The impact of distance and take-or-pay capacity reservations can potentially have a dramatic impact on wind energy. For example, based on an assumed postage-stamp price of 24 US$ /kW-yr (fairly typical for the USA) per zone, Stoft et al. (1997) calculate that a generator with a 100 per cent capacity factor and transmitting through only one zone would face a firm transmission cost of 0.27 US cents/kWh. In contrast, for a generator with a 20 per cent capacity factor and transmitting through three zones, the firm transmission price would rise to a crippling 4.11 US cents/kWh.

There are ways around the problem of take-or-pay firm contracts. The most obvious is to purchase non-firm transmission service. However, this involves certain complications. First, non-firm service would mean that the wind generator could be unable to access the transmission network during congested times, thus losing potentially significant revenue. Secondly, non-firm service also requires advance reservation. Even for hourly non-firm contracts, reservation is often required a day in advance. Thus, wind generators would face the same kind of risk as they face in the short-term generation markets, having to forecast their output a day in advance for any given hour. Because non-firm contracts are also take-or-pay, wind generators would continue to face the risk (albeit reduced) of over-reserving transmission capacity and leaving a certain amount unused or under-reserving transmission capacity and not being able to sell the full generated amount. If significant penalties are imposed on generators for not delivering the exact quantity reserved, this would cause a further problem. A third issue is that reliance on non-firm transmission increases the level of uncertainty and hence risk facing the wind plant. As discussed earlier in this chapter, increases in risk result in an increased cost of capital or, potentially, the unavailability of finance altogether.

The problem would be further reduced with the development of a robust secondary market for transmission rights. Such spot markets are beginning to develop in the USA, for example. If they allow trading sufficiently close to the time of actual use, then a wind generator with excess firm transmission contracts could re-sell some of its excess. With a strategic combination of firm, non-firm and secondary market contracts, a wind generator may be able to keep its transmission costs to a reasonable level. Nevertheless, intermittence would inevitably mean higher per-kWh transmission costs for a wind plant than a comparable fossil fuel plant.

Different proposals have been put forward to ease the impacts of transmission pricing on intermittent generation technologies, including allowing intermittent generators to use a 'pay-for-what-you-use' transmission tariff based on actual used capacity rather than reserved capacity (Ellison et al., 1997). Such tariff schemes could be accused of providing undue special treatment for renewables, however.

Stoft et al. (1997) argue that the entire system of take-or-pay capacity-based reservations is in need of fundamental re-thinking. Their argument can be summarised as follows. Transmission costs arise through two fundamental components: the fixed cost of building and maintaining the transmission network, and congestion charges arising during periods of high network use. Traditional capacity-based firm take-or-pay contracts address the issue of congestion by limiting the amount of generation capacity with access to the transmission network at any given time. Because non-firm generators are curtailed first during times of congestion, generators with firm contracts are assured access to the grid at a fixed price regardless of congestion. In other words, generators who sign firm transmission contracts are purchasing insurance against curtailment by pre-paying for the price of congestion. Thus, the cost of firm capacity reservation should equal the total expected costs arising in the system as a result of congestion.

The problem with the existing system, they claim, is that firm take-or-pay capacity reservations are also being used to pay for the fixed cost of building and maintaining the transmission system, which typically account for 80–90+ per cent of total transmission costs. Stoft et al., argue that this is economically inefficient and that congestion charges should be separated from fixed costs; the firm take-or-pay reservation should be an insurance charge which covers only the cost of congestion. The other 80+ per cent fixed costs are better addressed through a network access charge, which can be charged on the basis of actual energy transmitted rather than capacity reserved. This, they claim, would send the correct economic signals to all network users and result in the least-cost generation mix on the transmission system, while neither favouring nor discriminating against intermittent resources. Thus, with Stoft et al.'s recommended pricing system, only 10–20 per cent of transmission charges would be collected on a take-or-pay capacity basis, drastically reducing the penalty paid by wind energy plants for their intermittent nature.

In addition to transmission as discussed above, the electricity grid provides various other functions to maintain overall system stability, commonly known as 'ancillary services'. The US Federal Energy Regulatory Commission (FERC) defines ancillary services as including the following six categories: (1) scheduling, system control and dispatch service; (2) reactive supply and voltage control from generation sources service; (3) regulation and frequency response service; (4) energy imbalance service; (5) operating reserve – spinning reserve service; and (6) operating reserve – supplemental reserve service (Ellison et al., 1997).

With the development of competitive markets, these ancillary services are also being unbundled from transmission prices, and generators are beginning to have to pay for these services separately based on the level of ancillary services they require. Here also, wind power plants may be disadvantaged owing to their intermittent nature. In particular, wind plants could potentially require higher levels of regulation and frequency response service, energy imbalance service, spinning reserve service, and supplemental reserve service (Wind

Energy Weekly, 1997). The degree of ancillary services required will vary to some degree based on the wind technology in question. For example, stall-and-pitch regulated wind turbines would require reactive power, while variable-speed turbines would be capable of providing reactive power. On the other hand, stall-and-pitch regulated turbines would likely be better at providing frequency control than variable-speed turbines (Windpower Monthly, 1998c). Furthermore, the degree of ancillary services required would also depend on the particular rules in any given market. Ellison et al. (1997), for example, suggest that wind turbines should not be required to secure as much spinning reserve as fossil fuel plants because the aggregation of multiple turbines in a wind farm means that the risk of a wind farm going off-line is smaller than for a conventional plant of similar capacity operating on only one gas turbine or boiler.

Overall, pricing for transmission and ancillary services has the potential to create sizeable competitive disadvantages for wind energy, though many of the impacts can be mitigated. It is important that wind generators do not overlook these issues and also that regulators do not institute rules which unduly harm the prospects for intermittent resources' viability.

## Green marketing

The final issue examined in this chapter regarding the potential viability of wind energy in competitive power markets is that of green marketing. Green marketing is based on the premise that some customers will voluntarily pay extra to purchase electricity generated by renewable 'green' technologies. The environmental attributes of renewable energy are thus considered a value-added service which commands a price premium in the market place. In this sense, renewable energy is treated no differently than designer-label clothing, for example, for which people choose to pay extra even though the designer product is no more functionally useful than the less expensive non-designer product. Green marketing is thus a true market-based concept for environmental protection, in which people pay according to their own perception of the inherent value of clean power.

Because electricity markets involve the sale of a commodity product in which one electron is indistinguishable from another, electricity retailers in competitive markets must identify a strategy to differentiate themselves from rivals. Price is perhaps the primary differentiating factor between competing sellers, but environmental cleanliness can be a powerful marketing tool as well. Particularly in developed countries where electricity bills are a minor component of households' monthly expenditures, customers may well be willing to spend slightly more to purchase power from a provider whom they feel is more environmentally responsible. With wind energy being only slightly more expensive than fossil fuel-based electricity, there may be significant potential for wind to tap into this green market.

How large is this potential market? A large number of survey results in the USA indicate that between 40 and 70 per cent of respondents express a willingness to pay a premium in their electricity price for environmental protection or renewable energy (Farhar and Houston, 1996). Yet, actual US green marketing programmes implemented to date typically indicate a participation rate of below 2 per cent. The true potential of the green market is therefore very difficult to gauge. Most green marketing programmes are no more than two or three years old, and the diffusion rate of such programmes tends to be quite slow. Thus, over the next ten years, the size of green markets is likely to grow considerably compared to their current size. Nevertheless, it is widely acknowledged that survey responses significantly overstate consumers' true willingness to pay for renewable electricity. Rader and Short (1998), for example, suggest that the green market may never amount to more than a few per cent of the total electricity market.

Green marketing presents an array of both philosophical and practical sticking-points. From a philosophical perspective, critics of green marketing argue that it is not correct to ask a small percentage of the population to voluntarily pay a high price for renewable energy, since the benefits of their generosity will accrue to society at large. This is a classic 'free rider' problem. Individuals have an incentive to encourage others to participate but to avoid participating themselves. Considering that the environmental benefits of renewable energy accrue to all, it may be both more just and more effective to have all customers pay for renewable energy through a non-bypassable charge.

From a practical perspective, there are a variety of challenges. First and foremost, a credible disclosure and certification system is necessary to verify that marketers selling 'green' energy are truly generating with renewable resources. However, the challenges go deeper than this. Rader (1998), for example, has criticised California's green marketing programme as being largely a fraud, even when the 'green' electricity is in fact being generated by renewables. The reason for this claim is that most renewable electricity being sold in the California green market in 1998 was purchased from out-of-state utilities who were already recovering the cost of this renewable electricity from their own ratepayers. In other words, much of the renewable electricity being sold for a premium in California would have been generated anyway in another state, regardless of whether any customers participated in California's green market. As such, Rader claims that the California green market is not contributing to any net increase in US renewable energy generation and is merely creating increased profits for marketers. Others counter, however, that this is merely a transitional issue as the green market gets established, and that new renewable energy facilities (including wind plants) are in fact being built specifically to service the California green market.

The issues surrounding green marketing are therefore complex, and the long-term size and impact of the green market are unknown at this time. Nevertheless, it is a positive sign that there are several energy marketers for whom green energy constitutes their core strategy for attracting residential, commercial and wholesale customers in California.[11] Some of these programmes, such as Green Mountain Energy Resources' 'Wind For the Future' programme, specifically include new wind power development as part of their green market strategy. Over the long term, wind energy could benefit significantly from continued development of this competitive market to meet customers' desire for more environmentally benign electricity. Furthermore, California is by no means the only place with such programmes. Green marketing programmes are being tried throughout the USA as well as in the Netherlands, Australia and other countries. Green marketing programmes are discussed further in Chapter 7.

Overall, wind energy continues to face financing challenges when compared to conventional power plants. These challenges may increase with the coming of competition in generation markets and the decreasing availability of long-term fixed-price power purchase contracts. On the other hand, the advent of short-term forward markets, improved wind prediction techniques, potential $CO_2$ credit markets and green markets may all prove to be beneficial for wind energy's long-term viability. It is critical that wind energy generators learn to understand and function within the intricacies of these new markets.

# 6
# Environmental Considerations

Electricity generation is one of the world's most significant sources of air pollution. In the USA, for example, electricity generation accounted for 79 per cent of $SO_2$ emissions and 64 per cent of $NO_2$ emissions in 1998 (USEPA, 1998) and 35 per cent of $CO_2$ emissions in 1994 (USEIA, 1996). In addition, different electricity generation technologies can have a wide range of other environmental impacts including water pollution, radiation, flooding, visual intrusion and so on.

To the extent that such environmental impacts are regulated and controlled, the cost of meeting the regulations is incorporated into the cost of generating electricity from each power plant. Thus, for example, if regulations require that emissions of $SO_2$ be reduced by 90 per cent, the emissions control technologies necessary to achieve this are included in overall power plant costs and should be reflected by the price consumers pay for their electricity. However, the remaining 10 per cent of $SO_2$ produced would continue to be released to the environment and the cost of these remaining emissions would not be reflected in power prices but would instead be borne by the public at large, manifested as human health impacts, ecological damage and so on.

Such damages whose costs are borne by the public rather than by the buyers and sellers of electricity themselves are known as 'externalities' in the economic literature. More formally, externalities are defined as 'the costs and benefits which arise when the social or economic activities of one group of people have an impact on another and when the first group fail to fully account for their impacts' (ExternE, 1995). Chapter 5 explained that a proper economic analysis of power generation options should incorporate environmental externalities, since the costs of environmental damage are true resource costs borne by society.

In practice, however, environmental externalities are often ignored in electricity generation analyses. There are various reasons for this, including the fact that, historically, common resources such as air and water were considered 'free' and therefore available to be used (that is, polluted) by anyone in whatever manner they chose. Furthermore, the harm done by pollution is generally diffuse and thus often invisible, causing significant cumulative harm to society but going largely unnoticed (and therefore unopposed) by individuals. Even today, when the importance of environmental protection is well recognised, externalities continue to be largely ignored, perhaps primarily due to the difficulty of ascertaining their true monetary value. The economic analysis of wind energy in Chapter 4 also ignored environmental considerations and discussed the economic costs of generation technologies purely in terms of their more readily identifiable monetary costs such as capital cost, operations and maintenance, fuel and so on.

The propensity to ignore environmental considerations in economic analyses, as well as in financial analyses,[1] creates an advantage for highly polluting technologies at the expense of cleaner technologies. As wind energy is generally considered one of the most environmentally benign generation technologies, the failure to incorporate environmental factors in economic and financial analysis may create a key impediment to increased adoption of wind energy. This chapter therefore explores the environmental considerations of generation technologies in general and wind energy in particular. The chapter begins with a brief introduction to the valuation of environmental amenities, goes on to discuss some estimates of monetised externality values for electricity generation technologies and finally looks in detail at the specific environmental challenges surrounding wind energy.

## What is the environment worth?

Pollution affects people's well-being in a wide variety of ways. Air and water pollution from a power plant may cause health impacts which result not only in physical suffering but also in economic damage through increased health-care costs and reduced productive working days. Also, if pollution damages a stream or lake, for example, further economic damage may result in terms of reduced fish catch or reduced tourism. The economic cost of such impacts, while complex, can be estimated with relative ease and accuracy by examining actual health-care costs, salaries, tourism revenues and so on.

But how does one value less tangible costs? If people die prematurely as a result of pollution, how does one place a monetary value on these lives? Furthermore, the benefits of environmental amenities like forests are not reflected in *direct-use values* like tourism alone. Intangible *non-use* or *existence values* also must be considered. For example, how does one place a monetary value on the fact that people who never visit the Amazon rainforest may nevertheless obtain a certain satisfaction simply from knowing that the rainforest exists and continues to support vast wildlife and biodiversity? If people are willing to pay higher prices or forego a certain degree of economic growth in exchange for protecting the rainforest, then clearly such existence values are real and ignoring them results in sub-optimal development. The rainforest may also provide what are known as *option values*, such as the possibility that valuable medicines may be derived in the future from the rainforest and that destroying the rainforest today would eliminate that valuable future option.

Last, but not least, is the issue of global warming and the growing concern that human activities, particularly fossil fuel combustion, are permanently altering the earth's climate, with potentially enormous but unpredictable future worldwide impacts. If one has no idea what the size of future impacts will be but is fairly sure that the impacts will be negative, how does one account for such impacts in an economic analysis? The difficulty of placing a monetary value on such things and their inherently subjective nature, are the primary reasons why environmental considerations are often ignored in economic analyses. Ignoring these considerations, however, is equivalent to assigning them a value of zero, which is also clearly incorrect.

In reality, the environment is usually not ignored altogether, but is instead treated as a separate consideration outside of the economic analysis. The problem with this, however, is the difficulty which arises in comparing different options. For example, how does one choose between electricity generation option *A* which costs

0.05 US$/kWh and causes *X* amount of pollution resulting in 50 excess deaths per year, or option *B* which costs 0.04 US$/kWh and causes *Y* amount of pollution resulting in 80 excess deaths per year? When one chooses either option *A* or option *B*, one is making an implicit judgement about the value of human life in deciding whether it is worth spending an extra 0.01 US$/kWh to save 30 lives per year.

Therefore, in making particular technology choices, societies make implicit judgements about the value of unquantifiable factors such as human life or the existence value of a pristine wilderness area. Rather than leave such judgements implicit, it may be possible to make better-informed and more meaningful decisions if environmental amenities can be more explicitly and formally quantified; and a number of valuation techniques have been developed for this purpose.

## Damage costs

A widely used framework for valuing the environmental externalities of electric generation technologies involves defining a specific damage function associated with each type of environmental impact. Often called the impact pathway methodology, the four general steps of this approach are illustrated in Figure 6.1.

In Figure 6.1, a hypothetical power plant emits a certain level of particulate matter, whose dispersion into the atmosphere can be modelled with a dispersion model. The plant's location and the atmospheric dispersion pattern will result in a given exposure level in the population. Using an estimated dose-response function, this exposure level can then be translated into an impact level such as increased illness. Lastly, the monetary costs of these increased illnesses are calculated in terms of health-care costs, lost wages and so on. Thus, in this example, the four steps of the impact pathway approach allow the calculation of a direct monetary cost for damages resulting from particulate emissions of a given power plant.

One can see a number of difficulties with this approach. First, because pollutants' dispersion and contact with the population will depend on specific location, population density and atmospheric characteristics, a separate dispersion analysis must be carried out for each source of emissions, resulting in high analysis costs. Secondly, how does one establish the dose-response relationship between

**Figure 6.1** Impact pathway approach for development of environmental damage costs

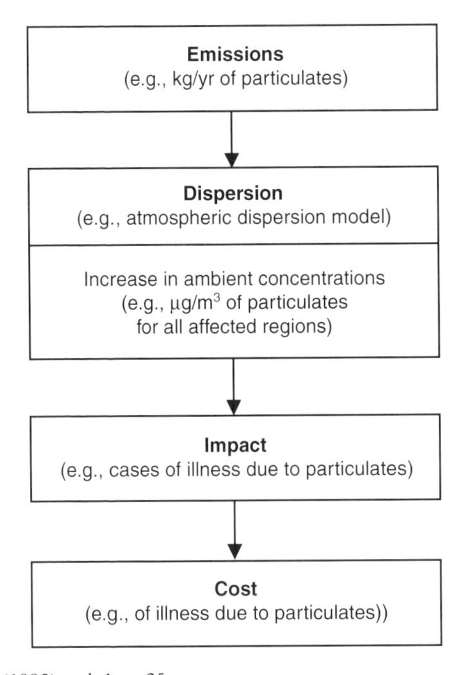


*Source*: ExternE (1995), vol. 1, p. 25.

exposure and illness? One might use epidemiological studies or controlled animal exposure studies, but neither of these are completely reliable and they too are expensive to carry out. Thirdly, and perhaps most controversially, how does one place a monetary value on the impacts, especially if the impacts involve intangible non-use values or option values?

The following paragraphs provide a brief introduction to some of the methods used for monetary valuation of environmental impacts when straightforward market prices (for example, of crop damage from pollution) are not available. These include hedonic pricing, travel costs and contingent valuation.

## Hedonic pricing

Hedonic pricing uses changes in the market value of related goods to infer the value of environmental amenities. For example, if a house located directly adjacent to a major highway costs less to buy than an identical house located 1 kilometre from the highway, then one might use the difference in price between the two houses to estimate the value of the environmental impact (air pollution, noise, visual impact) of the highway. In other words, hedonic pricing assumes that the prices of goods traded on the open market reveal the implicit value which people place on associated non-traded goods like environmental quality.

A widespread application of hedonic pricing has been to derive the value of noise pollution near airports. Because housing values tend to decline as houses get closer to airports and because this decline in value is assumed to be due to high noise levels, one can compare the change in noise level to the change in house prices to estimate the damage-cost function of noise. This is illustrated in Figure 6.2, where Function 1 represents a smooth linear function of declining house prices with increasing noise. In this case the slope of the line, or the unit change in price per unit change in noise, would represent the damage cost of the noise. However, the function may not be linear. It may be, for example, that there is a threshold level of noise up to which house prices show little sensi-

**Figure 6.2** Example of hedonic pricing to establish monetary damage cost of noise

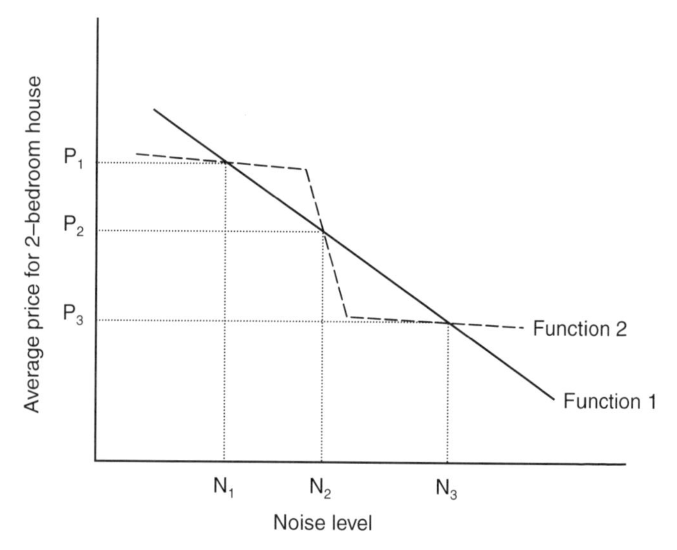

tivity and above which house prices decline drastically, only to stabilise again at a very high noise level. Function 2 in Figure 6.2 represents such a situation. In such a case, it would be more difficult to define the damage-cost function. However, both Functions 1 and 2 pass through the three points ($N_1$,$P_1$), ($N_2$,$P_2$) and ($N_3$,$P_3$). An analyst who had only these three data points would have no way of knowing whether the true damage cost was represented by Function 1 or Function 2.

Careful data analysis is therefore critical. Furthermore, the hedonic pricing method is only useful if the underlying market (for housing, in the case of Figure 6.2) is itself free of distortions. Price controls, housing segregation, lack of land availability, or any other number of factors could skew the relationship such that the price–noise function does not accurately reflect people's true preference level for quietude.

## Travel costs

The travel-cost method examines how much people pay to travel to a given site (for example, a national park) to determine a demand function for the site. This is illustrated in Figure 6.3, which plots the

**Figure 6.3** Travel-cost method for valuing environmental amenities

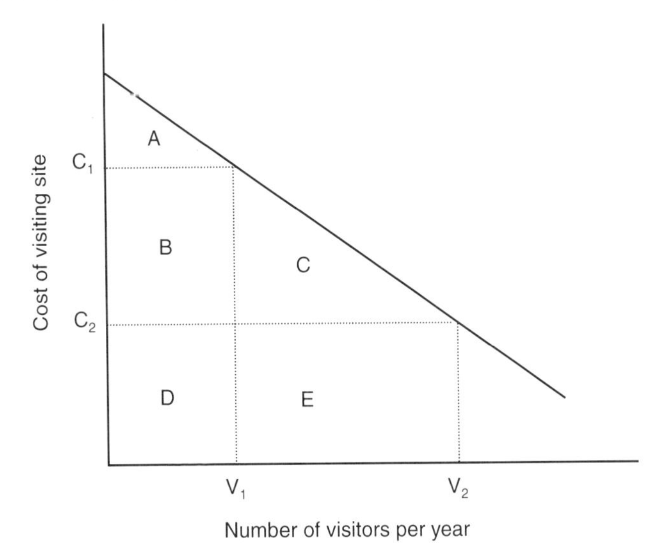

number of visitors to a hypothetical recreational site vs. the cost of visiting the site (transportation cost, entry fee, accommodation cost and so on).

Figure 6.3 shows an estimated demand function, in which the number of visitors is inversely related to the cost of the visit. When the cost is $C_1$, the number of visitors is $V_1$. This indicates that $V_1$ number of visitors is willing to pay *at least* $C_1$ to enjoy the site (a portion of them would have been willing to pay even more). Thus, area B plus area D represent the total cost which $V_1$ visitors spent in visiting the site. Area A represents what is known as the consumer surplus, or the additional amount that a portion of the $V_1$ visitors would have been willing to pay had it been necessary to do so. Thus, areas A plus B plus D represent the total amount that $V_1$ visitors were willing to pay to visit the site. Similarly, when the cost is $C_2$, the number of visitors is $V_2$. In this case, areas D plus E represent the total cost which $V_2$ visitors spend in visiting the site, and areas A plus B plus C represent the consumer surplus (Hakimian and Kula, 1995). Overall, the total area under the curve represents the minimum direct use-value of the recreational site.

The travel-cost method can also capture other factors such as the value of people's time to reach the site. Since this time does have some value (for example, in terms of lost wages), its value should also be included as part of the cost of visiting the site and hence as part of the direct-use value of the site. This method does not account for non-use values or the option value of preserving the site for the future, however.

### Contingent valuation

While the above-mentioned techniques are valuable in estimating the use values of many environmental amenities, they both suffer from an inability to deal with non-use or existence values such as, for example, the emotional well-being which people might derive from knowing that tigers or rhinoceroses continue to survive in the wild. Though these values are highly intangible, the fact that people give millions of dollars each year to wildlife preservation organisations clearly indicates that these existence values are real, since few of the people giving this money are likely personally ever to see the animals in the wild.

The contingent valuation (CV) approach differs from the above two methods in that it does not rely on observed market data to infer the value of environmental amenities. Rather, CV uses survey techniques to directly ask people how much they would be willing to pay to obtain a certain environmental improvement or what is the minimum payment they would be willing to accept in return for an environmental loss. Thus, a contingent valuation survey might include questions like 'how much would you be willing to pay in order to permanently set aside XYZ land as a nature preserve and prevent its future development?' The lack of reliance on observed behaviour is both the strength and weakness of CV.

Its strength lies in the fact that one can derive an estimated 'market' value for things for which no market exists. Thus, for example, CV was heavily used in trying to determine Exxon Corporation's liability for the damage caused to Alaskan wilderness from the Exxon Valdez oil tanker spill. Its weakness lies in the fact that the valuation is purely hypothetical and may thus be highly biased. Biases may be introduced, for example, by the way the questions are worded, by respondents' desire to influence the result, by respondents' lack of information about what they are asked to value and by the simple fact that people's hypothetical willingness to pay for things is different from what they will actually pay in reality. Contingent valuation is therefore perhaps the most versatile, as well as most controversial, method of environmental valuation.

**Other valuation issues**

All of the above methods have certain shortcomings, and in practice different methods may be used either in combination or as a means of establishing a range of estimated values. For example, in order to estimate the value of reducing deaths from environmental damage, one must estimate what is known as the 'value of a statistical life' (VSL), or the amount which a society is willing to pay to prevent the death of one average hypothetical person. This estimate might be derived using CV techniques, such as by asking people how much they would be willing to pay to reduce their likelihood of accidental death by a certain degree. If people were willing to pay an average of \$100 to reduce their risk of accidental death by 1 in 10 000, for example, then the estimated VSL would be \$100 × 10 000 = \$1 million. Or one might observe how much extra people are paid for dangerous occupations (for example, deep-sea diver, fire-fighter) in relation to their increased risk of death; this technique uses observed job market behaviour to estimate the VSL through people's willingness to accept payment for increased risk. Or, one might use people's voluntary expenditures on things which reduce the likelihood of accidental death, such as smoking cessation programmes or vehicle air bags (ExternE, 1995).

To estimate the value of morbidity (illness) impacts, one must estimate the value of lost time (including foregone earnings), decreases in well-being due to pain and suffering and costs of both averting and treating illness. The value of lost earnings and the cost of medical treatment are easy to estimate using observed wages and medical costs. To value pain and suffering, however, CV techniques are more useful. The values of all of these components are summed to obtain the overall morbidity value.

The issues involved in such valuation are very complex, such as the different valuation of voluntary vs. involuntary risk, the use of appropriate discount rates to value future costs and benefits and the potential for obtaining age-differentiated VSLs. The reader is asked to consult the references for treatment of such issues.

## Environmental damage costs of electricity generation

A number of studies have been carried out to try to estimate the value of environmental damages caused by electricity generation. Two of the most well known are the European Commission's 1995 ExternE *Externalities of Energy* study (ExternE, 1995) and the 1994 *New York State Environmental Externalities Cost Study* (RCG/Hagler Bailly, 1994).

The 1995 ExternE study represents one of the most comprehensive efforts to date to quantify monetary values of environmental externalities for a wide range of fuel cycles: coal, nuclear, oil, gas, hydro and wind. The study analyses full fuel cycles, from mining of fuel through power generation and waste disposal. The impacts analysed for the coal fuel cycle, for example, include damages relating to mortality, acute morbidity, chronic morbidity, occupational health, agriculture, forestry, aquatic impacts, materials impacts and noise.

Regarding wind energy, the ExternE study characterises the wind-energy fuel cycle as including the following environmental impacts: noise, visual intrusion, global warming, acidification, public accidents, occupational accidents, land use, bird mortality and radio interference. Though wind energy itself produces no air emissions which would result in global warming or acidification, construction and installation of the wind turbines does involve energy use which creates air emission impacts, though this depends on the nature of the energy resources already in use at the time of construction. Based on two wind farm sites in the UK, the ExternE study quantified the externalities of wind energy as shown in Table 6.1.

Noise values showed a wide variation depending on the population density surrounding the site. The study declined to place a specific value on the visual amenity due to lack of reliable studies and the great controversy surrounding the issue, particularly in the UK. The ExternE study estimated that visual impacts could range from less than 0.1 milli-ECUs per kWh (mECU/kWh) outside of designated scenic areas, up to 35 mECU/kWh in areas of major recreational importance.[2] Of the two specific sites analysed in the ExternE study, the study estimated the upper limit of the visual impact to be 1.9 mECU/kWh for one area with significant tourist traffic and 0.09 mECU/kWh for the other more typical UK site.

Land-use impacts of wind energy were deemed negligible because of the very small land area used by the actual turbines themselves and their compatibility with both agriculture and animal life. Bird mortality impacts were estimated to be negligible in the UK and throughout Europe, except in southern Spain where there is a high density of migratory birds. The ExternE report recommended continued study of this issue, however, acknowledging that a major study in California revealed significant mortality of raptors. But ExternE concluded that overall avian impacts in Europe were negligible as long as certain important bird sites were designated and excluded from wind farm development.

**Table 6.1** Estimated environmental externality values of wind-generated electricity

| *Category* | *External costs (mECU/kWh)* |
|---|---|
| Noise | 0.07–1.1 |
| Visual amenity | Not quantified |
| Global warming | 0.15 |
| Acidification | 0.7 |
| Public accidents | 0.09 |
| Occupational accidents | 0.26 |

*Source:* ExternE (1995), vol. 6, p. 118.

It should also be noted that, even in sensitive areas, many avian deaths may be avoided through improved siting and equipment selection. California's Altamont Pass, where a particularly large number of raptor deaths have occurred, is notable for being both the largest and oldest wind farm in the world. The Altamont site contains thousands of wind turbines, most of which are small (in the 100 kW range) by today's standards and thus not only spin faster but also cover a larger portion of the landscape than do larger modern turbines. Larger turbines with tubular towers (rather than lattice ones) are more visible, spin more slowly and are higher off the ground, all helping to avoid bird impacts. In addition, certain specific turbine locations within Altamont Pass appear to be responsible for the bulk of avian deaths in the area and the turbines at these most vulnerable sites are being removed as part of a repowering process to replace old small turbines with fewer new large ones (Wind Energy Weekly, 1998). It is therefore expected that avian deaths at Altamont Pass will decline in the future.

Looking at Table 6.1, if one were to assume a median noise value of 0.6 mECU/kWh and a median visual amenity value (of the two sites analysed) of 1.0 mECU/kWh, then summing the identified values in Table 6.1 would result in a total environmental externality value for wind energy of 2.8 mECU/kWh (0.0032 US$/kWh at the average 1997 exchange rate).

While this value is not trivial, it is less than one-tenth of the electricity generating cost. Furthermore, the global warming and acidification impacts listed in Table 6.1 are secondary impacts stemming from an assumption of fossil fuel-based primary energy use for turbine manufacturing. Though all fuel cycles (coal, nuclear, natural gas and so on) have such secondary impacts, the ExternE study included secondary emissions only for wind energy and did not analyse them for any of the other fuel cycles it studied. If therefore one were to exclude such secondary impacts for the purpose of comparison with other technologies, then the wind energy externality value would be only approximately 2 mECU/kWh (0.0023 US$/kWh).

A subsequent Danish study used the ExternE methodology to analyse the environmental externalities of both onshore and offshore wind farms in Denmark (Schleisner and Nielsen, 1997). Its results for wind energy are summarised in Table 6.2.

**Table 6.2** Danish ExternE national implementation: wind energy externality values

| *Impact category* | *Damage costs (mECU/kWh)* | | | |
|---|---|---|---|---|
| | *Offshore* | | *Onshore* | |
| Power generation | 0.01 | | 0.19 | |
| of which: Visual impact | | 0.00 | | 0.17 |
| Material production and manufacture | 0.66–3.64 | | 0.40–2.36 | |
| of which: Global warming | | 0.08–3.06 | | 0.06–2.02 |
| **Total** | **0.67–3.65** | | **0.59–2.55** | |

*Source*: Schleisner and Nielsen (1997), pp. 104–5.

The Danish study estimated total environmental externalities of the offshore wind farm to be 0.67–3.65 mECU/kWh (0.00076–0.0041 US$/kWh) and of the onshore wind farm to be 0.59–2.55 mECU/kWh (0.00067–0.0029 US$/kWh). The Danish study divided the damages into two broad categories: those which occur during power generation and those which occur during production and manufacture of the generating equipment and facilities. Those impacts occurring during power generation include accidents (to both the public and workers), noise, visual intrusion, bird impact, fish impact and interference with electromagnetic communication systems. Those impacts occurring during production and manufacturing are almost entirely air pollution impacts from fossil fuel use during manufacturing and installation of wind turbines. The wide range of estimated damage values from material production and manufacture reflect the significant uncertainty associated with global warming impacts.

For the offshore wind farm, virtually 100 per cent of the externalities were calculated to occur during the manufacturing and construction phases, due mostly to the global warming impacts of secondary $CO_2$ emissions. For the onshore wind farm also, secondary impacts of manufacturing were dominant. Of the 0.59–2.55 mECU/kWh of total onshore externalities, only 0.19 mECU/kWh were due to the power generation phase, of which 0.17 mECU/kWh were due to visual impacts and the remainder due mostly to noise. Thus, in comparison with the earlier 1995 ExternE study of the UK (highlighted in Table 6.1), the Danish study suggests a much greater impact from secondary $CO_2$ emissions and less impact from visual and noise considerations.

The 1994 New York State study (RCG/Hagler Bailly, 1994) did not consider secondary emissions during manufacturing. This study cited visual intrusion as the primary environmental impact of wind energy and estimated its damage value at 0–0.000018 US$/kWh in rural areas and 0–0.000939 US$/kWh in suburban areas. In other words, the maximum aesthetic damage value from wind turbines' visual impact was estimated to be slightly under one-tenth of one US cent per kWh. Other externality impacts of wind energy, including noise, land use, vegetation, wildlife and public and occupational safety, were not quantified in the study but were assumed to be negligible.

Comparisons between wind and other fuel cycles are difficult owing to differences in assumptions and methodologies. However, based on the other ExternE (1995) fuel cycle studies, Figure 6.4 provides approximate externality estimates for coal, oil, gas, nuclear and hydro in comparison to wind, in US dollars.[3] In addition, Figure

**Figure 6.4** Estimated total environmental externality ranges, by fuel type

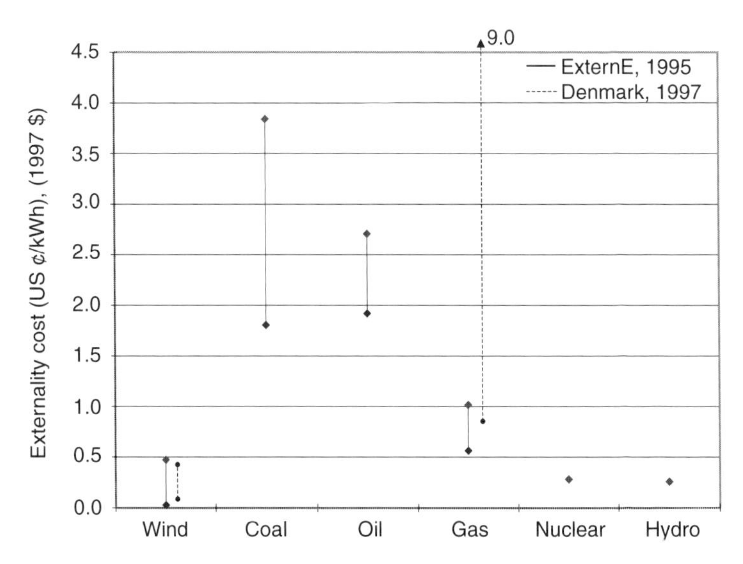


*Sources*: ExternE (1995) and Schleisner and Nielsen (1997).

6.4 also includes estimates for wind and gas from the 1997 Danish study (Schleisner and Nielsen, 1997).

The estimated externality values in Figure 6.4 are highly site-specific and the possible ranges on these values are very large. All numbers should be used only with extreme caution and a thorough understanding of the underlying assumptions. However, wind energy's environmental impacts are seen to be no higher than any other fuel and considerably lower than those of fossil fuels. Furthermore, while the estimated externality values for wind are broadly similar between the two studies, externality estimates for gas are considerably higher in the 1997 Danish study than in the 1995 ExternE study, due to different assumptions about global warming. In fact, the upper limit of the Danish estimate for gas is 9 US cents per kWh. Had the Danish study also analysed coal and oil, its externality estimates for these fuels would have been even higher. Such discrepancies between studies are common, especially given the enormous uncertainty associated with damages from global climate change.

Nevertheless, virtually all studies conclude that wind energy is one of the most environmentally benign electricity sources. Ignoring these environmental attributes in financial and economic analyses therefore results in a significant competitive disadvantage for wind energy. In addition, wind energy's environmental impacts are local, relatively predictable and primarily aesthetic, while those of fossil fuels and nuclear energy involve long-term risks whose magnitude cannot be accurately determined and which could potentially be much greater than the figures mentioned above.

## Social considerations

The previous paragraph highlights one of the paradoxes of the environmental debate surrounding wind energy. From a 'global policy' perspective, the local aesthetic impacts of wind energy appear more benign than the unquantified, long-term and large-scale impacts of things like global climate change and radioactive waste, whose overall impact on human health could be enormous. From a local perspective, however, the highly visible local intrusion of a wind farm may raise significantly greater passions than do abstract concerns of global long-term impacts. The result has been that wind energy facilities, generally acknowledged as one of the most environmentally benign electricity sources, have often had great difficulty in obtaining local planning permission for construction.

Public sentiment towards wind power development does not merely affect policy makers and planners; it affects the attitudes of investors as well. Gipe (1997) points out that bankers and investors take public opinion seriously in assessing projects' viability. The current state of the nuclear power industry clearly highlights this point. Negative public opinion towards nuclear power's environmental impact plays a key role around the world in deterring new investment in nuclear power, even though ExternE (see Figure 6.4) suggests that nuclear power's environmental impact may in fact be quite low.

Nowhere has the debate on the visual impacts of wind energy been carried out with more vigour than in the UK, where pro- and anti- forces have waged an acrimonious battle for public opinion. As a result, significant research into public attitudes towards wind energy has been carried out in the UK.

Surveys conducted in the UK show an overall positive public perception of wind energy and suggest that vocal opposition comes from a relatively small minority. Table 6.3 summarises a large number of UK polls regarding local public opinion towards specific wind farm projects.

These survey results show strong support for the various wind farm projects. Significantly, even when respondents expressed concerns prior to construction about wind turbines' potential intrusion into the local environment, surveys consistently found that respondents' impressions of wind energy improved once they had experienced wind farm operation for themselves, suggesting that the actual visual and noise impacts may be lower than commonly anticipated by the public.

For example, in the Delabole survey, the percentage who thought that wind turbines spoiled the scenery dropped from approximately 50 per cent before to 25 per cent after and those who thought wind turbines caused noise nuisance dropped from 86 per cent down to 20 per cent. The Bryn Titli project also shows significant improvement in public opinion subsequent to commencement of operations.

In general, therefore, even in the UK where public opinion has been divided over wind power, surveys consistently show strong

**Table 6.3** Summary of polls conducted in the UK regarding local public opinion towards wind energy projects

| *Location* | *Survey sponsor/organiser* | *Date* | *In favour (%)* | *Against (%)* | *Don't know (%)* |
|---|---|---|---|---|---|
| Delabole | Department of Trade and Industry | 1992/3 | 84 | 4 | 11 |
| Cemmaes, Powys | Department of Trade and Industry | 1992/3 | 86 | 1 | 13 |
| Llandinam, Powys | Countryside Council for Wales | 1992/3 | 83 | 3 | 14 |
| Llangwyryfon, Dyfed | Countryside Council for Wales | 1992/3 | 78 | 8 | 14 |
| Llandinam | BBC/University of Wales | 1994 | 76 | 17 | 8 |
| Rhyd-y-Groes | BBC/University of Wales | 1994 | 61 | 32 | 7 |
| Taff Ely | BBC/University of Wales | 1994 | 74 | 9 | 17 |
| Kirkby Moor, Cumbria | National Wind Power (NWP) | 1994 | 82 | 9 | 9 |
| Bryn Titli, Powys | NWP (preconstruction) | 1996 | 68 | 14 | 19 |
| Bryn Titli, Powys | NWP (open day) | 1996 | 94 | 3 | 3 |
| Trysglwyn, Anglesey | NWP (open day) | 1996 | 96 | 4 | – |
| Coal Clough, Lancashire | Liverpool University dissertation | 1996 | 96 | 4 | – |

*Source*: Anne Marie Simon Planning and Research (1996).

support, and the level of support increased as a result of actual first-hand experience with operating wind energy systems. Similar tendencies have been observed in Sweden (Hammarlund, 1996).

Nevertheless, concerns about the impacts of wind turbines on the local environment are very real and must be addressed squarely. This is particularly the case with large wind farms which, though economically more cost-effective, have greater visual and noise impacts than isolated single turbines. Openness and public involvement throughout the planning and siting process are critical ingredients in obtaining public consent and buy-in to wind power plants. Promotion schemes which encourage rapid wind energy development over a very short time-span can inadvertently result in public backlash by precluding public involvement due to compressed time schedules. This has been one of the main criticisms levelled at the UK's NFFO process (see Chapter 7), whose competitive and sporadic bidding process has encouraged the rapid development of wind farms in scenic areas, sometimes with only limited and belated local planning input.

Using both contingent valuation and hedonic pricing techniques, Danish surveys also indicate low overall levels of visual and noise disturbance in households located near windmills (0.0002–0.01 DKK/kWh, or 0.003–0.15 US cents/kWh at the average 1997 exchange rate) (Munksgaard et al., 1996), though some households consider the disturbance significant. Predictably, amongst people living in the vicinity of windmills, those who profit from the energy generated by their windmill co-operative consider the windmills as less of a nuisance than those living near windmills who receive no profit (Munksgaard et al., 1996). Allowing greater participation in the profits to those affected by the visual and noise impacts therefore helps diffuse the objections raised against wind energy.

Gipe (1995) highlights the need for local communities to perceive that they receive some of the benefits of wind power development and not just the costs. Close consultation and compensation from developers or opportunities for local populations to join wind energy co-operatives (as in Denmark, Germany and the Netherlands) may go a long way towards reducing public opposition. The visual landscape is, after all, public property; and the perception that 'outsider' developers profit while 'locals' pay the price of a landscape sullied with wind turbines is a sure-fire formula to foment public resistance. Reconciling the need for financial viability in an increasingly competitive electric generation industry with the local public's need for enfranchisement is likely to represent wind energy's key environmental challenge in the coming years.[4]

Job creation is another social consideration of great interest. Renewable energy systems, by virtue of their diffuse nature, have been touted as providing greater local employment than more centralised conventional electricity systems. Much of the research examining the linkage between renewable energy and employment was conducted in the late 1970s, and their conclusions suggested that wind energy (and renewable energy in general) is more labour-intensive and therefore provides higher employment than equivalent levels of conventional energy. Few such studies have been undertaken in recent years, but some recent results are summarised below.

In a survey of employment in the wind energy industry in the UK, Jenkins (1996) concluded that wind energy offers substantially higher employment opportunities than in the conventional power sector. This includes significantly higher local employment for operation and maintenance activities. This could provide the advantage of bringing employment into often economically depressed rural areas. On the other hand, where trained maintenance mechanics are in short supply, particularly in rural areas of developing countries, the dispersed nature of wind energy could cause maintenance difficulties and potentially result in lower reliability and higher costs.

In Denmark, an extensive macroeconomic analysis by the AKF Institute of Local Government Studies compared a wind-power-intensive scenario against a coal-based scenario and concluded that the difference in employment between the two was insignificant, though the wind scenario did result in slightly more jobs (Munksgaard et al., 1996).

Overall, it appears that wind energy's contribution to increased employment may be negligible to slightly positive. However, the issue of employment does not in itself provide strong justification for increased development of wind energy.

In absolute terms, the total number of jobs created by the wind energy industry is still small. Jenkins (1996) estimated that the UK wind energy industry employed roughly 1300 full-time-equivalent persons in 1994–5. In Denmark (the world's leading manufacturer of wind turbines), the Danish Wind Turbine Manufacturers

Association estimates that wind energy directly and indirectly provided roughly 9000 jobs in Denmark in 1995 and that worldwide employment in the wind energy industry is approximately 30 000 to 35 000 jobs (Vindmølleindustrien, 1996), based on 1200 MW of new installed wind capacity per year.

In developing countries, use of windmills can enhance local employment and improve countries' balance of payments by reducing the need for imported capital equipment and fuel, at least for low-technology water-pumping applications (Bhatia and Pereira, 1988). However, the employment and balance-of-payments implications of modern high-technology wind turbines for developing countries are less clear, as they will continue to require the import of equipment (though not fuel) as well as, perhaps, specialised labour.

# 7
# Wind Energy Policy

Great strides have been made over the last two decades in improving the technology, reliability, cost-effectiveness and overall understanding of wind energy. However, in spite of these improvements, significant barriers remain which must be overcome before wind energy can achieve substantial adoption within the general electricity market. These barriers have been discussed in previous chapters, but some of the most important are reiterated below:

- *Costs.* Wind energy technology costs have decreased significantly. In some cases, wind energy has become competitive with conventional sources, but in general, wind energy is still more expensive than conventional grid-based electricity generation. With the low natural gas prices which have prevailed over the last decade and the significant advances achieved in combustion turbine technology, full cost-competitiveness for wind energy remains elusive.
- *Dispatchability.* Because electricity cannot be readily stored, electricity generation output must be continuously increased and decreased to match supply with fluctuating demand on the electricity grid. The ability to control generation output (dispatchability) is thus a highly desirable trait for generation technologies. Wind energy resources are weather-dependent and inherently variable. The resulting lack of dispatchability increases the complexity of integrating wind energy into the grid, both in terms of physical grid operation and power sale contracts.
- *Small scale.* Conventional electricity generation technologies are typically over 100 MW in size and can reach well over 1000

MW. In contrast, wind energy technologies are small-scale, starting as small as under 1 kW for off-grid applications and increasing to perhaps 100 MW for a large wind farm. As a result, the transaction costs of planning, designing, building and operating wind energy facilities are typically much higher on a per-kW basis than those associated with conventional facilities. Small scale does offer some advantages for wind energy as well, such as dispersed modular implementation and reduced transmission and distribution investments, but these benefits are not always accounted for in economic calculations.

- *Environment.* Wind energy causes fewer overall negative impacts on the environment than conventional energy sources. However, these advantages are often ignored by decision makers when comparing wind plants with conventional power plants. On the other hand, due to its distributed local nature, wind energy can have local environmental impacts such as visual intrusion and noise; and these have made installation difficult in some areas.

- *Institutional bias.* Utilities typically have very limited experience of wind energy and tend to regard renewable resources with suspicion, particularly given the above existing barriers. This conservatism often results in wind energy being shunned even when it is attractive from both a technological and economic standpoint.

As a result of such barriers, special policies have been and continue to be necessary for wind energy to penetrate the electricity market. Some policy mechanisms, such as environmental taxation, aim to correct existing market failures by recognising technologies' differing environmental impacts and taxing them accordingly. Other mechanisms, such as investment subsidies, aim to expand the market size and thereby stimulate technological advance, economies of scale and overall cost reduction, with the eventual goal of eliminating the need for such subsidies. The various mechanisms are not mutually exclusive and are often used in combination. This chapter provides a description of policy mechanisms which have been used by countries to promote renewable energy in general and wind energy in particular. The chapter is divided into two parts, the first describing generic policy mechanisms, and the second describing specific countries' policies and experiences in more detail.

## Power purchase agreements

Reliable power purchase contracts are perhaps the single most critical requirement of a successful renewable energy project. The majority of renewable energy projects have been implemented by independent developers unaffiliated with utilities. The only possibility for such facilities to sell their power is to have access to the utility's transmission and distribution grid and to obtain a contract to sell power either to the utility or to a third party by wheeling through the utility grid. Because renewable energy projects are generally considered risky by financial institutions, a reliable, stable long-term revenue stream is extremely important for obtaining finance at a reasonable cost, as discussed in Chapter 5. Creation of reliable markets for independent power has thus been the cornerstone of essentially every successful renewable energy strategy. The most famous example of this is perhaps the 1978 PURPA law in the USA, which mandated that utilities purchase all independently generated power at their avoided cost; but other countries such as the UK, Denmark, Germany and India have all developed explicit (but differing) rules providing guaranteed power purchase agreements for renewable electricity.

However, mandating that utilities purchase power at their avoided cost is not in itself sufficient for successfully promoting renewable energy; determining an appropriate level of avoided costs is similarly important. Avoided-costs are calculated based on the marginal generation unit whose costs the utility could avoid by purchasing the renewable energy in question. While the concept is straightforward, calculation of avoided costs is complex, particularly when they must be forecast many years into the future. As a result, avoided cost calculations can vary significantly depending on the assumptions used. If the calculated avoided costs are not sufficiently high, wind energy projects may remain unable to compete against conventional sources, and further incentives may be necessary.

In addition, utilities are often reluctant to purchase independent power in spite of regulatory mandates to do so, and ways of overcoming this intransigence may also be necessary. Furthermore, the entire concept of 'avoided costs' becomes nebulous as generation markets are deregulated and move towards a competitive footing. As described in Chapter 5, deregulation of electricity markets has the potential to greatly increase the challenge for wind power in obtaining long-term power purchase contracts.

## Investment incentives

Investment incentives are often used to reduce project developers' capital costs and thus induce developers to invest in renewable energy. Incentives are typically paid either by the government through the general tax base or by utility customers through a surcharge on their utility bills. They can take a variety of forms, but some of the most common are described below.

### Investment subsidies

Direct capital investment subsidies can be provided per kW of rated capacity or as a percentage of total investment cost. Such direct subsidies are the most straightforward incentive and are attractive for their simplicity, but they must be strictly monitored against abuse and to ensure that project costs are not artificially inflated. A capable and vigilant regulator is thus essential in order for subsidy funds to be efficiently allocated. Germany and Finland are among the countries which offer direct subsidies for renewable energy investments. Other countries, such as Sweden and the Netherlands, provided such subsidies in the past but phased them out. Sweden reintroduced investment subsidies in 1998.

### Investment tax credits

Investment tax credits are similar to investment subsidies and serve to lower capital costs by allowing plant owners to reduce their taxes by the amount invested in qualifying projects. They can be useful in enticing profitable enterprises or high-income individuals to enter the renewable energy market to reduce their tax liabilities, but they can be inefficient if investors are more interested in maximising their tax shelter than in achieving actual electricity production. Investment tax credits are less transparent than direct investment subsidies, which may improve the political acceptability of tax credits but also increases their complexity and reduces their effectiveness. The most famous – and infamous – use of investment tax credits was in the USA to stimulate wind energy development in the 1980s. The strategy played a major role in the creation of the modern wind energy industry but also suffered widespread abuse and created a political backlash still felt to this day. Another drawback of tax credits is that small project developers may not have sufficient pre-tax income to fully absorb the tax credits (Wiser and Pickle, 1997a), thus limiting the range of investors who can benefit from such policies.

### Other investment tax incentives

A wide variety of other investment tax incentives exist. For example, import duty exemptions or reductions have been used in developing countries such as India and China to lower the cost of imported equipment. Other tax incentives include accelerated equipment depreciation, property tax reductions, and value-added tax (VAT) rebates. Such mechanisms can be used to lower projects' capital costs, though, as with all investment incentives, there is a danger that some of the incentive will be captured by equipment vendors through higher prices. Again, tax incentives can be politically expedient, as it is usually easier for governments to avoid collecting taxes through tax credits than to collect the taxes and then disburse them as explicit subsidies. But from a public policy standpoint, such expediency must be carefully balanced against the complexity and distortions inherent in manipulating the tax system.

### Preferential finance

The cost of raising capital is a major factor in all investment projects. This is particularly the case for infrastructure projects like power generation which involve large up-front costs, and long construction lead times and operating lifetimes. Thus, improved financing terms such as lowered interest rates or longer repayment horizons can significantly reduce project costs. Governments such as Germany and India have created special financing agencies to provide loans for renewable energy projects at below-market interest rates. Furthermore, many development organisations, including the World Bank, provide loan guarantees which reduce risks for commercial lenders and thus improve commercial loan terms and availability.

## Production incentives

Like capital investment incentives, production incentives are subsidies to reduce the cost of producing electricity from renewable sources. As with investment incentives, production incentives can be paid from the general tax base or through a surcharge on customer utility bills. However, unlike investment incentives, which are paid based on initial capital costs, production incentives are paid per kWh of electricity generated. Production incentives can be superior to investment incentives by eliminating the temptation to inflate initial project costs and by encouraging developers to build reliable facilities which maximise energy production. The shift from investment incentives to production incentives in the USA was clearly influenced by this concern and by the abuses encountered by early investment incentive schemes.

However, production incentives also suffer from one clear disadvantage compared to investment incentives. Because production incentives are paid per kWh generated, project developers and funders must rely on the assumption that the incentives will continue to be available in future years. Elimination of production incentives due to policy changes, government budget cutbacks or political whim can have devastating financial impacts on renewable energy projects. By contrast, investment incentives which are paid up-front are not subject to changing political forces once the incentive is paid. On the other hand, investment subsidies can also be subject to political uncertainty at the time of construction, as evidenced by the USA's year-to-year extension of its investment tax credit in the late 1980s and early 1990s, subject to yearly Congressional approval, which ultimately led to the bankruptcy of LUZ International, the world's most successful solar thermal electric power developer (Wiser and Pickle, 1997a). Nevertheless, for developers, investment incentives are generally much safer against political risk than production incentives.

### Per-kWh production subsidies

Production incentives can take different forms, the simplest being the direct cash subsidy, paid per kWh of electricity produced. Countries using such subsidies include the UK, Denmark and Germany. However, the level of subsidy can be determined in a variety of ways. In the UK, the level is determined through a competitive auction, while in Denmark and Germany the level is administratively set as a percentage of the residential electricity tariff. In California, under its electric-utility industry restructuring law, existing renewable electricity projects are paid an administratively determined production incentive, while new projects must competitively bid for the per-kWh incentive.

### Per-kWh production tax credit

As with capital investment incentives discussed above, production incentives can also be provided as tax credits rather than as direct subsidies. This has been the strategy employed by the USA since 1992, for example, in promoting wind and biomass energy. Production tax incentives are subject to the same advantages and disadvantages (compared to production subsidies) as were described above for investment incentives. The advantages appear to be primarily those of political expediency, while disadvantages include complexity and lack of ability of certain parties to fully absorb the tax credit. Furthermore, Kahn (1996) has argued that tax credits' usefulness is limited because, in order to take full advantage of tax credits, projects must be financed with a greater proportion of high-cost equity and a lower proportion of low-cost debt than would otherwise be the case.

## Renewables set-aside

A renewables set-aside mandates that a certain percentage of total electricity generated comes from renewable sources, and reserves a specific portion of the market exclusively for renewables. A set-aside policy thus recognises that renewable technologies may not be able to compete on the open market and instead creates a separate market within which renewable projects must compete amongst themselves. Thus, such policies rely on market forces and competition to stimulate cost reductions and further renewable technology development.

Though open competition among all renewable technologies in one reserved market may be theoretically appealing, in reality different renewable technologies are in widely differing states of development. Some technologies, such as landfill gas or waste incineration (assuming they are considered 'renewable' at all), are highly developed and can virtually compete in the open electricity market;

others, such as geothermal and wind, can in some cases compete on the open market; and others such as photovoltaics are rarely competitive. Thus, open competition even amongst only renewable technologies would still result in a few technologies dominating this market. Therefore, if diversity of technologies is desired, it may be necessary to further allocate the renewables market into specific percentages for specific technologies.

The Non Fossil Fuel Obligation (NFFO) system in the UK is the most famous of such set-aside schemes and divides the renewables market into several technology bands. For each technology, power purchase contracts are awarded by the electricity regulator on a competitive basis, thus relying on market forces within each technology band. A similar but even more market-oriented concept involves tradable renewable energy credits and includes the Netherlands' Green Labels programme, as well as the Renewables Portfolio Standard (RPS) being attempted in some states in the USA. The NFFO, Green Labels and RPS are described in greater detail later in this chapter under the UK, Netherlands and USA sections.

## Externality adders

As traditional energy planning has largely ignored the environmental externalities of power production, this has favoured technologies with high environmental impacts and discriminated against more environmentally benign technologies. Some regulators have attempted to address this issue by increasing the hypothetical cost of conventional power plants through an environmental externality charge or 'adder' in the planning stage. Such adders can improve the likelihood of renewable energy plants being built by increasing the apparent cost of conventional technologies. Typically, externality adders are included only in the planning stage for resource selection but are not actually charged on operations, thus not affecting power plant dispatch once projects are built. Some US states have used externality adders for power project planning.

## Environmental taxation

Like the externality adder, environmental taxation adds to the cost of fossil fuel-based energy by imposing a per-kWh tax on the basis of pollutant emissions. Environmental taxation can thus provide a competitive advantage to renewable technologies with low emissions. Unlike the externality adder, however, environmental taxes involve actual payment of money and are not merely a hypothetical charge for planning purposes only. Current debate regarding global climate change resulting from $CO_2$ emissions has stimulated much interest in the idea of carbon taxes, but actual implementation of carbon taxes to date has been largely limited to northern European countries, including Denmark, Finland, Netherlands, Norway and Sweden (EEA, 1996). However, taxes on other emissions, including sulphur oxides and oxides of nitrogen ($SO_x$ and $NO_x$) are more common.

Care is necessary in determining how taxes are calculated. For example, carbon taxes on biomass energy would be complicated by the fact that biomass emits significant carbon when burned, but over its lifetime creates zero net carbon emissions. Thus no net emissions would occur from sustainably harvested biomass, but net emissions would occur if biomass is harvested through deforestation. Environmental taxes have different impacts on different renewable energy technologies. Non-emitting technologies like solar or wind benefit from all environmental taxes, but biomass could be hurt by taxes on $NO_x$ or particulates, for example. On the other hand, if burning biomass for electricity reduces uncontrolled burning of biomass waste products in the field, then biomass electricity would actually reduce overall emissions of $NO_x$ and particulates.

## Research, development and demonstration grants

The mechanisms outlined above can all be used to enhance current implementation of commercial renewable energy projects. In addition, other incentives can be used to improve the general technological and knowledge base necessary for more long-term stimulation of renewables. In particular, many governments provide research, development and demonstration (RD&D) grants for renewable energy technologies, as well as for resource assessment, environmental considerations and other related areas. According to the International Energy Agency, OECD spending on renewable energy research and development (R&D) was on the order of US$880 million in 1995, the largest percentage of which came from the USA, followed by Japan, Germany and Spain (IEA, 1997b).

Different countries' R&D programmes have focused on different renewable technologies. The USA's programme is fairly evenly distributed, though solar technologies (solar heating and cooling, photovoltaics and solar thermal electricity) account for over 50 per cent of total funds. Japan and Germany also place their largest emphasis on solar photovoltaics, though the non-solar component is dominated by geothermal in Japan and wind in Germany. Sweden and Canada's R&D programmes are dominated by biomass, while Denmark's R&D spending is split roughly equally between biomass and wind.

However, unlike with other more direct incentive mechanisms, spending on R&D does not necessarily translate into a high level of installed renewable capacity. For example, between 1973 and 1988, the USA and Germany spent roughly US$380 million and US$79 million, respectively, on wind energy R&D, but Denmark came to dominate the world wind-turbine manufacturing market, spending only US$15 million on R&D during the same period (Righter, 1996). R&D spending must be carefully integrated with reliable long-term markets if R&D is to translate into practical application. The same drawback can be observed in the current UK programme, in which there is little overlap between the technologies targeted by R&D spending and those supported through renewables set-asides (NFFO).

## Government-assisted business development

In addition to providing RD&D assistance, governments can also indirectly stimulate the implementation of renewable energy by providing various types of business development assistance. Possible types of assistance include encouraging the formation of risk-sharing consortia, providing technology export promotion, setting technical and safety standards and providing certification, and others.

One mechanism successfully employed in Sweden is known as 'technology procurement', in which the government organises a consortium of buyers (for example, of wind turbines), specifies technical specifications, and solicits bids from manufacturers. The consortium guarantees a minimum amount of purchases to the manufacturer who can meet the specifications at the lowest cost, thus reducing technology development risks for manufacturers while ensuring high quality at low price for the purchasers.

Government export-promotion assistance includes agencies such as the US Export–Import Bank, the US Overseas Private Investment Corporation, the Export–Import Bank of Japan, and Germany's Kreditanstalt für Wiederaufbau.

## Green marketing

Green marketing is a relatively new concept in which electricity customers are given the option to voluntarily pay a higher price for electricity generated from renewable sources. This concept stems from the fact that surveys conducted in many developed countries indicate that people would be willing to pay a price premium for clean energy; and green marketing thus allows people to 'vote with their wallet' for renewables. As the ultimate 'market-driven' approach to environmental protection, green marketing is likely to receive increased emphasis in liberalised electricity markets. And in fact, as one of the few non-price means of distinguishing one's service in a commodity market, green energy could well become a major marketing strategy for energy companies in the competitive electricity market. However, if green marketing is to be successful in promoting renewables, it requires a well-informed, environmentally-motivated public that is willing to pay extra for a diffuse and intangible benefit.

Green marketing programmes have been very popular in the Netherlands, where utilities have in some cases had difficulties in keeping up with demand. Other countries experimenting with green marketing include the USA and Australia. Green marketing is expected to play a major role in the restructured competitive US electricity industry, but the US restructuring process is still too new for any conclusions to be drawn. Green marketing is also discussed at the end of Chapter 5.

## Tradable $CO_2$ credits

The 1992 United Nations Framework Convention on Climate Change (UNFCCC) and its subsequent 1997 Kyoto Protocol require nations to reduce their emissions of greenhouse gases, of which $CO_2$ is the most prominent. A variety of mechanisms are being discussed to help achieve global $CO_2$ reductions at the lowest overall cost.

One such mechanism is the joint implementation of projects between countries, often funded by one country and implemented in another, in which the participating countries share credit for the achieved emission reductions. Another proposed mechanism involves tradable $CO_2$ emission permits, in which countries can meet their emission reduction requirements either by reducing their own emissions or by purchasing emission permits from other countries who are able to reduce their own emissions more cheaply and sell their excess permits.

These mechanisms are all still under consideration, and it is not yet known how they will work in practice. Nevertheless, any binding commitments on the part of countries to reduce their $CO_2$ emissions will lead to a de facto $CO_2$ credit market in which projects which reduce $CO_2$ emissions (like wind energy) will receive some form of financial compensation. In fact, renewables market set-aside programmes which involve tradable renewable energy credits, like the Dutch Green Labels programme and the proposed US Renewables Portfolio Standard, are essentially no different from tradable $CO_2$ credit markets. Over the long term, such $CO_2$ credits may become one of the driving forces of renewable energy investment as the UNFCCC becomes fully implemented.

## Other policy mechanisms

Other mechanisms exist for promoting the implementation of renewable energy. Two such mechanisms which allow flexible access to the electricity grid are described below.

### Wheeling

In some cases, an electricity consumer may wish to self-generate using renewables, but the location of the renewable resource (for example, wind or biomass) may be different from the location of the consumer, requiring some transmission capability. Or in other cases, a large customer may wish to purchase its power directly from a private (perhaps renewable) generator, located off-site, to avoid purchasing from the local utility. In either case, such arrangements would not be feasible unless the utility's transmission grid can be used to transmit, or 'wheel', the power from the generation site to the consumer's site. Wheeling provisions can be implemented to allow such private transmission over utility lines by paying a charge to the utility. Such wheeling provisions for renewable energy have been implemented in India, for example.

The ultimate manifestation of this is known as 'retail wheeling', in which all electricity consumers can freely choose to purchase power from any electricity supplier through a bilateral contract, and the transmission system operator and distribution system operator are merely paid a per-kWh fee for operating their lines and maintaining system reliability. Such competitive systems have been implemented in Norway and Sweden, for example, and are currently being introduced in the USA. This ties in closely with the green marketing concept described above and can allow any customer to choose to purchase renewable energy directly from any supplier without contracting through the utility.

### Electricity banking (net metering)

Renewable energy sources such as solar and wind are variable by nature and thus cannot be relied upon to produce electricity at the precise time of need. To overcome this drawback, electricity banking is a contractual system in which renewable generators can essentially 'store' their electricity in the utility grid, to be used later. This amounts to the ability to sell one's generated power to the utility at a certain price and then purchase the same amount of power back from the utility at a later date for the same price plus payment of a service fee. Electricity banking can be particularly useful for season ally variable resources such as solar, wind and run-of-river hydro. For example, a self-generator using run-of-river hydro may find that his power production is far greater than his consumption during the wet season but is too low during the dry season. Through electricity banking, this customer could then pay the utility a service fee to act as a bank, absorbing the excess power in one season and delivering it back to the customer in another season. Electricity banking has been implemented in India.

## Country experiences with grid-connected renewable energy policy

The renewable energy policy mechanisms described above are rarely implemented in isolation. Rather, countries typically follow a multi-pronged approach incorporating various mechanisms. In some cases, this is a result of a clearly developed strategy and a recognition that any one mechanism may not be sufficient to achieve the desired implementation rates. In other cases, the use of multiple mechanisms may merely be the result of poor policy co-ordination and a piecemeal approach. Significant insight can be gained by examining countries' policies for renewable energy promotion and analysing their successes and failures. The remainder of this chapter takes a closer look at the policies for grid-connected renewable energy in seven countries: USA, UK, Netherlands, Denmark, Germany, India and Sweden. These seven countries have been at the forefront of wind energy development over the past two decades.

## America

The USA, and particularly the state of California, has been the site of some of the greatest renewable-energy policy successes as well as failures over the last 20 years and offers many valuable lessons. The following pages highlight the various forms of renewable energy promotion carried out in the USA, including the PURPA law, tax incentives, the California system benefits charge, green marketing, the renewables portfolio standard and others.

### PURPA

The birth of the US renewable energy industry and the independent power industry in general can essentially be traced to the passage of the Public Utility Regulatory Policies Act (PURPA) of 1978. This law mandated that utilities purchase all power generated by 'qualifying facilities' at their 'avoided cost'. Qualifying facilities (QFs) include cogeneration plants, and electricity plants of less than 80 MW capacity fuelled by renewable sources and less than 50 per cent-owned by electric utilities or their affiliates (Gilbert, 1991). 'Avoided costs' refer to those costs which the utility would otherwise have to pay to generate the electricity itself. At the time of PURPA's passage, virtually no non-utility power generation existed, and few people foresaw the enormous growth in non-utility generation which would occur over the next decade and which would permanently change the electric utility industry.

Though PURPA was a federal law, actual implementation of the law was left to individual states; and different states acted with varying levels of interest. Among those who implemented PURPA most aggressively were the states of California, Texas and Louisiana. Aggressive implementation of PURPA in California was due to serious power shortages stemming from significant delays in the construction of three utility nuclear power plants (Hamrin and Rader, 1992), and the state's desire to promote resource diversity through small power plants and renewables. However, in spite of power shortages and the PURPA mandate, utilities were reluctant to sign contracts with independent generators which would end the utilities' monopoly of the generation market and which would make them reliant on untested suppliers. To overcome such utility reluctance and to smooth the contractual process in general, the California Public Utilities Commission (CPUC) ordered the institution of standard contracts, known as 'Standard Offers' (SO), to be signed between utilities and QFs.

The standard offer contracts included four contract types, known as SO1, SO2, SO3 and SO4, and were the key to California's PURPA implementation. SO1 and SO3 contracts pay for energy and capacity on an as-available basis, while SO2 contracts pay for energy on an as-available basis but pay fixed capacity prices for up to 30 years (CPUC, 1993). The most popular and most controversial contracts, the SO4, evolved through various forms but essentially provide fixed payments for both energy and capacity (see Mead and Denning, 1991). Utilities were required to sign standard offer contracts with all sellers who met the necessary criteria.

The interim SO4 (ISO4) contract, available from 1983 to 1985, resulted in phenomenal QF activity. Though conventional wisdom expected no more than 1000 MW of ISO4 contracts, by 1985 more than 15 000 MW of ISO4 contracts had been signed, leading to fears of over-capacity and forcing the CPUC to suspend the ISO4 within a mere two years.

Though the bulk of QF projects have been fossil fuel-based cogeneration projects, a large number of renewable energy projects were built as well, many with SO4 contracts. The California experience with PURPA offers many valuable lessons, as outlined below.

### *Contracts*

Early experience with PURPA indicated that merely requiring utilities to purchase non-utility power at their avoided cost was insufficient. Two additional barriers had to be surmounted: (1) utility reluctance to sign contracts, and (2) complexity, delays and high transaction costs of independently negotiated contracts. The creation of standard contracts was essential in overcoming these dual barriers. The four different standard contract types offered in California allowed significant streamlining of the contractual process while still providing sufficient flexibility to meet the needs of diverse actors in the market. For example, the option of front-loading payments to renewable projects helped ease debt repayment by matching projects' revenue streams to renewable facilities' cost streams.

*Avoided costs*

PURPA mandated that non-utility power be purchased at the utility's avoided costs, but how to correctly calculate avoided costs was not adequately resolved, including for example, whether they included only short-run marginal costs or long-run marginal costs. In trying to stimulate QF development, the CPUC was, in retrospect, too generous in setting avoided costs, though this was not evident at the time. The greatest problems occurred with the ISO4 contracts, which set fixed energy and capacity payments based on projected future avoided costs. In the early 1980s, energy prices (and hence avoided costs) were expected to continue rising inexorably, and thus prices paid to QFs in ISO4 contracts were locked in at an ever-increasing rate.

For example, ISO4 prices paid by Southern California Edison and by Pacific Gas & Electric were both set to rise from under 6 cents/kWh in 1983 to over 12 cents/kWh in 1997 (CPUC, 1993). In reality, however, the utilities' actual avoided costs dropped over that time period from approximately 5–6 cents/kWh to approximately 3 cents/kWh. This discrepancy allowed very large profits for the QFs at the expense of utility ratepayers. However, had the utilities not signed such QF contracts, their alternative options at the time were to build very costly nuclear and coal plants; so, in fact, the rate impacts of the above-market QF contracts are often overstated compared to the utilities' favoured alternatives at the time. Nevertheless, determining an appropriate level of avoided costs is essential for a successful independent power programme.

*Fixed vs. variable payments*

Though the fixed avoided-cost ISO4 contracts turned out to be expensive in retrospect, the fact that they guaranteed a fixed revenue stream greatly increased the QFs' financial security, which was in turn reflected to some degree in lowered finance costs and hence lowered power plant costs. And, in fact, for renewable power plants which were (and often still are) considered highly risky, it may not have been possible for many renewable plants to be financed at all were it not for the security offered by the fixed payment streams. This therefore highlights the trade-off between fixed payments and variable payments. Fixed payments are more secure and are thus more successful at stimulating power plant development, particularly for technologies considered to be risky. Variable payments which fluctuate (in line with current gas or oil prices, for example) entail substantially greater risk for the developer and thus may prevent many projects from ever being built, but they do help to avoid windfall profits (either for the utility or the independent developer) as a result of fluctuating avoided costs, as occurred under the ISO4.

*Unlimited contracts vs. bidding*

When the standard offer contracts were first proposed, no maximum amount of contracted QF capacity was specified, as the level of activity was expected to be small. By the time the need for a cap on capacity was recognised, far more QF contracts had been signed than the CPUC had intended, in large part due to the generous terms offered. This danger of oversubscription can be avoided through a bidding process which restricts the maximum number of contracts to be signed if available supply exceeds demand. This was the direction pursued in California for the Final Standard Offer 4 (FSO4) contract, as a successor to the ISO4.

## Tax incentives

In addition to PURPA, tax incentives have been the other driving force of renewable energy development in the USA. The following is a list of some of the most important tax incentives available at various times for renewable energy.

*Federal tax credits and depreciation allowances*

The Energy Tax Act of 1978 provided a business tax credit of 15 per cent for certain energy technology investments, including many renewable technologies like wind power. These credits were in place through 1985. A generic business investment tax credit (ITC) of

10 per cent was also in force at the time and was available until 1986. Furthermore, a five-year accelerated depreciation of investments was allowed through the accelerated cost recovery system (ACRS) established as part of the Economic Recovery Act of 1981. The ITC and ACRS were available for all types of investments (not just energy and not just renewables) but were particularly valuable for renewable energy projects entailing high capital costs (Cox et. al., 1991). Other tax credits were in force for customer-sited renewables such as photovoltaics (PVs) and solar hot-water systems. Most tax advantages for renewables were eliminated by the 1986 Tax Reform Act, but the Energy Policy Act of 1992 instituted a 1.5 cent/kWh production tax credit for wind and closed-loop biomass, and permanently extended the 10 per cent business energy investment tax credit for non-utility investment in solar and geothermal facilities (Wiser and Pickle, 1997a). The production tax credit expired in June 1999 but, as of the time of this writing, was widely expected to be renewed by the US Congress.

*State tax incentives: California*

Various states also provided tax incentives in addition to those provided by the federal government. In California, the available tax incentives for renewables included a 25 per cent energy investment tax credit which was available through 1995, was reduced to 15 per cent in 1996, and expired at the end of 1996. Accelerated depreciation for state tax purposes was also available.

As a result of these federal and state tax advantages, during the mid-1980s an investor in a California wind energy plant could recover 60 to 80 per cent of his investment entirely through tax advantages, even if the power plant never generated any electricity (Cox et. al. 1991), and in some cases the tax write-offs could be as high as 90 per cent of the investment (Righter, 1996). Predictably, the result of such generous tax benefits was mixed. Combined with the generous power purchase contracts available under PURPA, investment in renewable energy projects could be highly profitable. But with many projects (particularly wind) being developed primarily for tax shelter purposes, project performance in terms of electricity generation was often far below expectations. It was to avoid such abuses that tax incentives were changed from capital cost-based tax credits to production-based credits in 1992.

### Other government incentive programmes

The federal and various other state governments also provided other assistance for renewable power projects. These included loans and loan guarantees from the federal Small Business Administration, state sales tax exemptions, local property tax reductions, and special technical assistance under programmes like the Wind Energy Systems Act of 1980.

### Research and development

This chapter does not look in detail at countries' renewable energy research and development programmes, as they are not directly related to establishing working renewable energy projects. However, a few observations are worthwhile regarding the USA's renewable energy R&D programme. Over the last 20 years, US government expenditures on renewable energy R&D (expressed in terms of 1991 dollars) have varied between a high of approximately $900 million in 1980 and a low of slightly over $100 million in 1990 (EIA, ~1992). In 1995, the US government's renewable energy R&D expenditure was $393 million (in 1995 dollars), allocated approximately 54 per cent to solar, 15 per cent to biomass, 12 per cent to wind and 10 per cent to geothermal (IEA, 1997b). Despite the USA's very large investment in renewable energy technology R&D, very little actual implementation of new renewable energy projects has taken place during the last decade. And as outlined earlier in this chapter, the USA's large investment in wind energy R&D has not translated into a successful commercial wind turbine industry, particularly in comparison to Denmark's low R&D spending but high commercial success. This demonstrates that R&D programmes on their own are generally of limited value in creating successful renewable energy projects, unless they are combined with more market-oriented support.

### California renewables system-benefits charge

The mechanisms described above mostly reflect past policies which were responsible for the creation of the US (and particularly Californian) renewable energy industry. After the end of generous tax credits and the suspension of new California long-term PURPA contracts in the mid-1980s, US renewable energy activity declined significantly and has remained at a low level for the past decade.

However, with the advent of electricity industry restructuring over the past few years and the arrival of full retail-level competition in California, significant new attention is again being turned towards renewable energy. As with PURPA and with tax incentives, California has again taken the nationwide lead in terms of restructuring its electricity industry. This section highlights some of the new policy developments in California and shows how they affect the renewable energy market.

The key reform in California has been the advent of retail competition such that, as of March 1998, all electricity customers are free to choose their electricity supplier. Utilities have lost their retail monopolies and must compete against a wide array of energy providers, including other utilities, to sell electricity to their customers. California utilities have substantially been divested of their generation assets, and generation contracts are now structured either as bilateral direct-access contracts or as sales to the California Power Exchange spot market. Utilities continue to own their transmission grids but have transferred control to the California Independent System Operator who manages the entire state's transmission grid and power plant dispatch. Utilities continue to own and operate their local distribution grids but must provide non-discriminatory access to any competing electricity retailer.

In terms of renewable energy, there has been significant concern that, without some form of continued government-mandated funding, the entire established California renewable energy industry may not survive in the new competitive market. As a result, a new renewables support mechanism has been adopted to collect a total of $540 million from electricity customers between 1998 and 2002 to support existing, new and emerging renewable electricity generation technologies (California Assembly Bill, no. 1890, Ch. 854, Sec. 381, 1996). These funds are collected by the utilities through a non-bypassable charge on distribution service (often called a 'system benefits charge'). Allocation of these funds to individual projects has been made the responsibility of the California Energy Commission (CEC).

The CEC has divided the funds into four primary categories: existing technologies (projects operational before 23 September 1996), new technologies (projects operational after 23 September 1996), emerging technologies and consumer credits. The allocation of funds to these four categories has been established as follows (CEC, 1997):

- **Existing technologies.** The existing technology funds provide support to already existing projects which continue to require financial support to remain operational. The existing technologies are further divided into three tiers, in which Tier 1 (currently least cost-effective technologies) includes biomass and solar thermal projects, Tier 2 includes wind and Tier 3 (currently most cost-effective) includes geothermal, small hydro, digester gas, landfill gas and municipal solid waste. For the existing technologies, incentives are paid on a per-kWh production basis, and the amount is determined by the lesser of (a) the administratively determined target price minus the market clearing price, or (b) available funds divided by generation, or (c) specified production-incentive caps. The target price is set highest for Tier 1 (5 cents/kWh in 1998 declining to 3.5 cents/kWh in 2001), while the Tier 2 and Tier 3 target prices are 3.5 cents/kWh and 3.0 cents/kWh, respectively. Furthermore, the production-incentive cap for all tiers is 1.0 cent/kWh except for Tier 1 in 1998–9, for which the cap is 1.5 cents/kWh.

  The CEC provides the following example for how to calculate the incentive for existing projects (CEC, 1997, pp. 29–30):

  > Assuming that the total level of generation by certified solid-fuel biomass and solar thermal suppliers during a monthly payment period is 300 GWh, the available funds during that period are $3 million, and average market clearing price levels are 3.2 cents/kWh, the results of the three tests described above (for 1998) will be as follows. (1) Target price minus market clearing price levels equals 5.0 cents/kWh minus 3.2 cents/kWh = 1.8 cents/kWh; (2) Available funds divided by eligible generation equals $3 million ÷ 300 million kWh = 1.0 cent/kWh; (3) The production incentive cap is 1.5 cents/kWh. Based on the lesser of these three calculations, determined in this case by available funds divided by generation, the production incentive for technologies in Tier 1 would be set at 1.0 cents/kWh for that month.

  These subsidies for existing projects disappear after the year 2001, requiring all existing technology projects to survive within the open market. Note also that any repowered PURPA projects holding SO2 or SO4 contracts are also classified as existing technologies.

- **New technologies.** For new technologies (projects operational on or after 23 September 1996), all technologies are treated within the same category, and funds are allocated based on a simple auction, with funds allocated to those projects requiring the least support. In other words, higher-cost technologies like solar or biomass do not receive any preferential treatment over cheaper technologies like digester gas, in the case of new technologies. Investors are thus expected to invest in the most cost-effective technologies as dictated by the market, with no technological preference indicated by the state. Production incentives are subject to a maximum cap of 1.5 cents/kWh and will be awarded to the lowest-cost bidders up to the point where funds are exhausted. For projects awarded incentives, these incentives are to be paid out over a five-year period subsequent to project commissioning.

- **Emerging technologies.** 'Emerging technologies' are classified to include photovoltaics, solar thermal electricity, small wind turbines of 10 kW or less, and fuel cells using renewable fuels. Funds for this category are distributed on a project-by-project basis through issuance of specific requests for proposals. Forms of assistance are flexible, based on the needs of the individual projects, and could include, for example, consumer financing assistance, loan guarantees or interest-rate buydowns, per-kWh production incentives, or capital-cost buydowns.

- **Consumer-side account.** The fourth category, consumer credits, are meant to help stimulate an active 'green' retail market in which consumers choose to purchase electricity from renewable energy suppliers. Consumers who choose such green power can receive an incentive applied to their electricity bills which is determined by the lesser of (a) available funds divided by eligible renewable generation, or (b) a 1.5 cent/kWh incentive cap. The green electricity market is described further below.

The CEC's distribution allocation is based on the need to support technologies with widely differing characteristics and levels of maturity, and to keep the renewable energy industry's existing projects alive while stimulating new additional developments. How successful the strategy will be is unclear, as its implementation is just beginning. However, the level and duration of funding do raise concerns. The Union of Concerned Scientists has argued that the overall $540 million funding level will be insufficient to maintain the present aggregated level of non-hydro renewables in California (CPUC, 1997). Though nothing prevents funding levels from being increased in the future, there is no current sign of this happening; and in any event the uncertainty over future funding beyond 2001 is likely to place great strain on financing any new projects. Furthermore, the CEC's guidelines for new projects stipulate that new projects will receive funding for only five years after commissioning. However, the UK's early experience with the NFFO (discussed later in this chapter) showed that contracts of even seven years were too short to obtain reasonably priced finance for projects. In other words, the short funding period provided by the California legislation (to stimulate a rapid transition to a fully competitive market) may in itself prevent renewable energy projects from developing sufficiently to become competitive.

**Green marketing**

In conjunction with and in addition to the system benefits charge-based funding described above, the other interesting (and possibly more important) development in California is the emergence of the 'green' power market. Green marketing allows consumers to voluntarily choose to pay higher electricity prices to ensure that their electricity is generated using renewable energy technologies. In the competitive California retail electricity market, environmental friendliness could potentially become one of the major marketing tools for electricity retailers, particularly for serving residential and small commercial customers whose energy consumption is relatively low and not very sensitive to energy prices. As of late 1999, green power was being offered in the residential market by several companies, including Cleen 'n Green Energy (Preferred Energy Services), Commonwealth Energy, Edison Source, Green Mountain Energy Resources, Keystone Energy Services, New West Energy, PG&E Energy Services and the Sacramento Municipal Utilities District, while others such as the Environmental Resources Trust, Foresight

Energy Company and the Automated Power Exchange Green Power Market had established green power services for the wholesale market (EDF, 1999).

Beyond California, several utilities around the USA are implementing green marketing programmes for wind energy. These include Public Service Company of Colorado, Central and Southwest Corporation in Texas, Fort Collins Lighting & Power in Colorado, Dakota Electric in Minnesota and Traverse City Light and Power in Michigan (Wind Energy Weekly, 1996–1997). Many utilities have also offered green marketing programmes for photovoltaics (Wiser and Pickle, 1997b). In general, such green marketing programmes have so far been modest; and though some programmes have been enthusiastically received and others are still just getting started, overall customers do not appear to be joining green marketing programmes at the high rate indicated by responses to surveys of their willingness to pay for renewable energy. Such programmes may therefore still require more time and publicity before beginning to have a real impact; but increased marketing associated with liberalisation in states like California could greatly increase this momentum.

**Renewables portfolio standard**

In addition to the California system benefits charge-based renewables programme and green marketing campaigns described above, the third support mechanism for renewables receiving attention in the USA is the Renewables Portfolio Standard (RPS). Under the RPS, all retail power suppliers would be required to obtain a certain minimum percentage (for example, 10 per cent) of their electricity from renewable energy, in the form of 'renewable energy credits' (RECs). An REC would be a type of tradable credit representing one kWh of electricity generated by renewables. Electricity retailers could obtain RECs in three ways. (1) They could own their own renewable energy generation, and each kWh generated by these plants would represent one REC. (2) They could purchase renewable energy from a separate renewable energy generator, hence obtaining one REC for each kWh of renewable electricity they purchase. Or (3) they could purchase RECs, without purchasing the actual power, from a broker who facilitates trades between various buyers and sellers. In other words, RECs are certificates of proof that one kWh of electricity has been generated by renewables, and these RECs can be traded independently of the power itself. The basic idea of the RPS is both to ensure that a certain minimum percentage of electricity is generated by renewables and to encourage maximum efficiency by allowing the market to determine the most cost-effective solution for each electricity retailer: whether to own renewable generation, purchase renewable electricity, or buy credits, and what type of renewable technology to use (Rader, 1996).

The idea for trading RECs is based on the emissions trading concept used in the 1990 Clean Air Act Amendments in which total national sulphur emissions are capped, and emission permits are issued to allocate the total allowed emissions amongst polluters. Those who are able to reduce their sulphur emissions cheaply can do so and sell their excess emission permits, while those for whom emission reductions are costly can avoid reducing emissions by purchasing excess permits from others, thus encouraging the most cost-effective overall emission reductions. With the RPS, because each REC would represent one kWh of electricity generated somewhere with renewables, an electricity retailer who purchases RECs from a broker without actually purchasing renewable electricity would still be ensuring that the renewable electricity is generated somewhere within the state or country. Though the RPS was considered and ultimately rejected in California in favour or the system benefits charge system described above, various versions of the RPS have been approved by state legislatures and/or public utility commissions in several US states including Maine, Nevada, Massachusetts, Vermont and Arizona (solar only) (Rader, 1997; Windpower Monthly, 1998d). Several federal utility-restructuring bills under consideration by the US Congress also include provisions for an RPS.

## USA lessons learned

The experience with promoting renewable energy in the USA provides a wide variety of lessons, which are summarised here.

### *PURPA*

As outlined earlier in the discussion on PURPA, the PURPA experience highlighted: (a) the importance of providing reliable power purchase contracts which provide a predictable revenue stream; (b) the importance of establishing appropriate avoided costs as a means of setting contract prices; (c) the trade-off between providing stable fixed-price contracts and more flexible variable-price contracts; and (d) the benefits of bidding or some other mechanism to restrict total capacity and encourage cost reduction.

*Tax credits*

The success of US renewable energy tax credits has been decidedly mixed. In retrospect, it can be said that generous capital cost-based tax credits reduced the incentive for developers to build reliable projects and in some cases encouraged outright fraud. On the other hand, the generous tax credits were in many ways responsible for the creation of the modern wind energy industry. Without such generous incentives, it is unclear whether investors would have chosen to invest in such risky and untried technologies. Nevertheless, as the industry has matured, there is now little need for the level of incentives provided by the USA in the early years. The shift to production-based tax credits and the much more limited scope of current credits reflects a shift towards greater emphasis on cost-effectiveness and reliable production. Where capital cost-based incentives are provided, they must be carefully monitored by an effective regulator.

*Policy stability*

One of the chief drawbacks of US renewable energy policy has been its continuously shifting nature, varying between over-generous incentives and virtually no incentives. Such boom-and-bust cycles encourage speculation by short-term profit-seekers and do little to promote a sustainable cost-effective renewable energy industry. The suspension of SO2 and SO4 PURPA contracts and the elimination of tax credits in the mid-1980s led to widespread bankruptcies and contributed to the loss of much valuable experience. It is therefore essential that incentives and policies be modest and stable, with emphasis on long-term development.

The clearest example of the need for policy stability is the experience of LUZ International, the world's leading developer of parabolic-trough solar thermal power plants during the 1980s. Following the elimination of federal and California renewable-energy tax credits in 1986, the US Congress extended the 10 per cent federal-investment tax credit on a temporary year-to-year basis, but the tax credit's existence could not be assumed beyond any given year.

Furthermore, California maintained a 25 per cent investment tax credit which would only be effective as long as the federal tax credit remained in place. Thus LUZ, which was building one solar thermal electric facility in California per year, was required to obtain the site licence, raise capital and build the entire power plant within one year to ensure the availability of the tax credits for each project. Such year-to-year uncertainty significantly raised LUZ's costs of building projects and also made the company highly vulnerable to changes in policy. In 1989, amid great uncertainty regarding extension of tax credits and resulting cost overruns, investors lost confidence and began backing out, leading LUZ into bankruptcy (Wiser and Pickle, 1997a). Again, years of experience, investment and expertise were needlessly lost in this process as a result of policy instability; and development of solar thermal electricity has been stalled ever since. Similarly, in mid-1999, the US Congress allowed the 1.5 cent/kWh wind energy-production tax credit to expire. Although eventual renewal of the production tax credit was widely expected, nonetheless, the uncertainty surrounding this caused another boom-and-bust cycle. Late 1998 and early 1999 witnessed a dramatic jump in new wind power projects as developers rushed to commission their projects before the tax credit's expiration. This was followed by a major drop in activity and job losses in mid-1999 as new wind plant orders evaporated following the expiration.

*Research and development*

Though the USA has invested vast sums of money in renewable energy R&D over the years, its failure to provide stable and reliable markets for renewable energy has meant that R&D expenditure has not translated into operating commercial projects. R&D must be co-ordinated with appropriate market-stimulation policies.

*Electric industry restructuring*

Significant thought has been applied to renewable energy's fate in restructured competitive markets. However, in spite of this, California's renewable energy policy for its new market has been greeted with scepticism for several reasons. First, incentives for new projects are to be paid for only five years, far too short to effectively reduce project risks and lower financing costs. Second, beyond the four-year 'transition period' from 1998 to 2001, no further support is envisioned for renewables, which will be entirely dependent on the 'green' market to compete against conventional power plants. Given the green market's highly uncertain size and stability, little renewable energy development can take place unless project developers are able to take on large amounts of risk on their own balance sheet. Notably lacking in all restructuring efforts has been the implementation of a stable contractual mechanism to ensure long-term sustained development.

## UK

The UK has promoted renewable energy technologies through its Non Fossil Fuel Obligation (NFFO), first introduced in 1990 subsequent to the privatisation of its electricity supply industry. The NFFO was originally created as a support scheme for the country's existing nuclear power plants, which could not otherwise survive in the new competitive electricity market; but the NFFO has also emerged as a powerful mechanism for promoting renewable energy. The renewables NFFO sets aside a certain portion of the electricity market to be supplied by designated renewable energy technologies. Within each technology band (wind, biomass, landfill gas and so on), developers submit bids of proposed projects; and the projects with the lowest per-kWh bid price are awarded power purchase contracts. Regional electricity companies (RECs) are mandated to purchase power from NFFO-awarded renewable electricity generators in their service area at the premium price determined through the bidding process. The RECs are reimbursed for the difference between the NFFO premium price and their average monthly power-pool purchasing price through the Fossil Fuel Levy which is collected from all electricity consumers (Mitchell, 1995).

Though it has not been without its faults or controversies, the NFFO is to date the most famous and most successful example of a 'market-oriented' competition-based approach to renewable energy promotion and incorporates many of the most important lessons highlighted in the discussion of US experience above. In particular, the NFFO heeds the following three lessons:

- Some level of *contract stability* is necessary to attract finance for risky capital-intensive projects like renewable energy which produce electricity at above-market costs.

- Some form of competition, bidding or ratcheting-down of incentives is useful for stimulating technological innovation and cost reduction.

- Production incentives paid on a per-kWh basis are more effective and cost-effective than capital cost-based incentives.

The NFFO is implemented through periodic auctions, of which five had been carried out as of 1998. The NFFO has not only been successful at stimulating substantial numbers of renewable energy projects, but its competitive process has also stimulated rapid reductions in the electricity price demanded by projects. Though the exact nature of the allowable technologies and contracting structure has changed over time, Table 7.1 summarises the number of projects established and the drop in prices achieved between 1990 and 1997.

As shown, the NFFO has been successful in stimulating a significant amount of renewable energy development at reasonable cost and in creating a viable renewable energy industry where none previously existed. The cost of wind energy declined from a highest-awarded bid price of 10 UK pence/kWh in 1990 down to an average bid price for large projects of 3.53 pence/kWh in 1997. And by the 1998 NFFO round 5, the average price for large wind projects further declined to 2.88 pence/kWh (BWEA, 1999). However, Mitchell argues that the particular competitive system used in awarding contracts has been bureaucratic and in some cases expensive, compared to the non-competitive systems used in Denmark, the Netherlands and Germany (Mitchell, 1995). Therefore, though one cannot necessarily equate 'competitive' with 'cheap', the NFFO does demonstrate one of the few successful working models for providing stable renewable energy contracts within a privatised competitive electricity-industry structure.

The following highlight some of the key lessons learned through the NFFO process (Mitchell, 1995):

- In early NFFO rounds, contract duration was limited to end in 1998, resulting in very short contracts which greatly raised the cost of finance. Starting with the NFFO round 3 in 1994, contract lengths were increased to 15 years, significantly easing financing terms. In general, however, small renewable energy projects have

**Table 7.1** UK Non Fossil Fuel Obligation (NFFO) status between 1990 and 1997

| *Technology* | *NFFO 1 (1990) highest awarded price (pence/kWh)* | *NFFO 4 (1997) Average bid price (pence/kWh)* | *Contracted projects as of December 1999* | | *Commissioned projects as of December 1996 Capacity (MW)* |
|---|---|---|---|---|---|
| | | | *Number* | *Capacity (MW)* | |
| Biomass (energy crops and agricultural and forest waste) | 6.0 (anaerobic digestion) | 5.17 (anaerobic digestion) –5.51 (gasification) | 22 | 197 | 0 |
| Small hydro | 7.5 | 4.25 | 84 | 50 | 20 |
| Landfill gas | 6.4 | 3.01 | 167 | 344 | 119 |
| Municipal and industrial waste | 6.0 (mass burn) | 2.75 (fluidised bed) –3.23 (CHP) | 50 | 795 | 99 |
| Wind | 10 | 3.53 (large) –4.57 (small) | 178 | 603 | 71 |
| Other | | | 39 | 109 | 91 |

*Source*: Mitchell (1997) Tables 2, 3. (As of November 1998, 1 US$ = approximately 0.60 UK pounds.)

had difficulties in attracting finance even in spite of the NFFO. A successful financing mechanism is therefore critical, especially for small projects.

- The highly competitive nature of the NFFO, its stop-start auction process, and the initial limitation of premium payments up to 1998 required projects to be developed very quickly with limited public planning input. In the case of wind energy, such development occurred at sites with very high wind speeds, often located in scenic areas. The lack of co-ordination between the NFFO and local planning procedures has been significantly responsible for the well-publicised local backlash against visually intrusive wind power developments. The competitive process also has favoured more visually intrusive (but more cost-effective) wind farm projects over the single-turbine individually or co-operatively owned projects frequently found in Denmark, which are typically much more acceptable to local communities.

- There has been limited overlap between the technologies supported by the NFFO and those supported through government R&D funding. As a result, few of those technologies supported in the R&D stage have found subsequent commercial markets. Better co-ordination of R&D and market support programmes like the NFFO could result in more effective government spending on renewables.

- Vast oversubscription of the NFFO auctions has meant that many developers who prepared projects did not ultimately secure contracts, resulting in significant uncertainty, wasted effort and hardship on the emerging industry.

- Despite its success in lowering renewable energy prices, a transition strategy has still not been identified for many renewable energy technologies to move from the protected NFFO world to the competitive open market. How renewable energy will fare once NFFO subsidies end is still not clear. Technologies like waste-to-energy and in some cases wind may be approaching viability in the open market, but others such as biomass are still far from achieving commercial competitiveness. Other technologies such as PV have not yet reached the stage of even being cost-

> effective enough to qualify for the NFFO. Concerns about the ultimate transition to the open market has also served to highlight the need for clear and fair pricing in all areas of the electricity grid, including appropriate compensation to locally generated renewable electricity for avoiding transmission and distribution requirements.

As the premium contracts signed for renewables under NFFO rounds 1 and 2 expired at the end of 1998, the issue of how such projects will make the transition to the open market is a very real one. One strategy being pursued is for renewable generators to pool their projects in the Renewable Generators Consortium to negotiate collectively with retail electricity suppliers selling to the 'green' market (Windpower Monthly, 1998d). In the UK, as in California, full retail-level competition was being introduced in 1998, stimulating interest by electricity retailers to differentiate their product by offering 'environmentally friendly' electricity services. How the green market will evolve in the UK is not yet clear, but it is also raising the issue of how competitive green marketing on the open market should coexist with the protected NFFO market. Renewable projects holding current NFFO contracts are required to sell their output to the local regional electricity company and may not sell to any other retailer even if a higher price is available. Thus, green marketers can have difficulties finding sufficient available suppliers of green power even if they are able to attract sufficient purchasers. Such issues are yet to be addressed.

## The Netherlands

The Netherlands has traditionally been one of the leaders in renewable energy development, with particular emphasis on wind energy. Through the 1980s the Netherlands ranked third in the world in installed wind capacity after the USA and Denmark. Support for wind energy in the Netherlands has included both R&D grants (starting in the 1970s) and a variety of market-stimulation mechanisms. Initial market-stimulation programmes in the 1980s included the Integrated Programme Wind Energy (IPW), which provided a subsidy of 35–40 per cent of investment costs for newly built turbines, and the 'MilieuPremie' environmental bonus from the

Ministry of Housing, Physical Planning and Environmental Management (VROM) which provided capital subsidies for wind turbines in selected suitable areas and a bonus for low-noise turbines. The IPW was replaced in 1990 by the support programme 'Application of Wind Energy in the Netherlands' (TWIN), which provided further subsidies for technology development and market stimulation and which lasted until 1996 (Wolsink, 1996).

In addition, in 1990 the Dutch government set new goals to reduce $CO_2$ emissions and encouraged utilities to introduce Environmental Action Plans (MAPs) to invest in energy conservation, renewables and $CO_2$ reduction. The MAPs were funded by a wires charge on distributed electricity and stimulated new wind-capacity development, though efforts were hampered by utilities' difficulties in securing adequate sites and their reluctance to purchase power from independent generators at a sufficiently high buy-back tariff (Wolsink, 1996).

Dutch renewable energy policy has been significantly modified during the last few years and is substantially affected by liberalisation of the Dutch electric utility industry. Direct subsidy programmes such as the TWIN were eliminated in 1996, but several other more market-oriented mechanisms have been put in place to encourage the development of an 'environmentally conscious' economy. These mechanisms include those listed below (Kwant, 1996; Novem, 1998).

### Green funds

Since January 1995, several banks have been offering 'green funds' in which the public can invest at an average interest rate of approximately 4 per cent per year. This interest is free of income tax for investors, allowing banks to pay a lower interest rate to investors than for other investments. In return, the green funds are obliged to invest a minimum of 70 per cent of their capital in 'green projects', which include most renewable energy technologies, including wind. Project developers must apply for 'green certification' from the Ministry of VROM before they can access capital from green funds, which can be borrowed more cheaply than standard loans due to their tax-free status. As of late 1996, the public had invested NLG 900 million in green funds, and these investments are primarily supporting wind energy and district heating projects.

### Accelerated depreciation

The Accelerated Depreciation on Environmental Investments Scheme (VAMIL) was introduced in September 1991, and, as the name suggests, allows environmental investments (including wind energy technologies) to be depreciated more rapidly than under normal depreciation rules, thus reducing taxable income during projects' initial years and improving cash flow.

### Regulating Energy Tax

This tax was introduced in 1994 and is imposed on households and small businesses for electricity and natural gas consumption when their consumption rises above a certain minimum level. The tax, as of 1998, kicks in only for electricity consumption in excess of 800 kWh per year and gas consumption in excess of 800 $m^3$ per year, so highly efficient consumers can avoid paying the tax. The tax amounts to an approximate increase of 15 per cent in electricity prices and 25 per cent on gas, though these are planned to be increased in the future. In addition to encouraging energy conservation, the tax also supports renewable energy because the Regulating Energy Tax collected on electricity generated by renewables is paid directly to the renewable generator as an incentive, rather than to the government.

### Green electricity

This concept is identical to the green marketing concept outlined in the discussion on the USA earlier in this chapter, in which customers voluntarily choose to pay a higher price for electricity generated by renewable technologies. In late 1996, the price premium for green electricity was on the order of 0.04–0.08 NLG/kWh above the average standard tariff of 0.285 NLG/kWh (as of November 1998, 1 US$ = approximately 1.9 NLG). The World Nature Fund monitors and certifies the renewable energy content of the green electricity schemes. The Dutch government is also proposing to reduce the value-added tax rate for green electricity from 17.5 per cent down to 6 per cent to help offset the price premium paid by the consumer.

### Green labels

Perhaps the most interesting development is that, as part of the liberalisation process and as part of the MAP 2000 covenant between the government and utilities to increase renewable electricity's market share, a tradable 'green labels' market has started, as of January 1998 (Windpower Monthly, 1997). Under current laws, local energy distribution companies (LEDCs) must purchase renewable electricity from independent power generators at a price based on the current wholesale pool price of electricity (currently around 0.08 NLG/kWh) and the Regulating Energy Tax refund (approximately 0.03 NLG/kWh). However, under the new programme, in addition, the LEDCs must issue green labels to the renewable generator, based on the number of renewable kWh sold to the grid (one green label represents 10 000 kWh of renewable electricity). The renewable generator can then sell these green labels on an open market to distribution utilities who will all be required to own a certain quota of green labels as part of their agreement with the government. With wind energy, for example, given current production costs of approximately 0.16 NLG/kWh and current payments from utilities of approximately 0.11 NLG/kWh (0.08 pool price plus 0.03 Regulating Energy Tax refund), the renewable generator would have to sell its green labels for at least 0.05 NLG/kWh to realise a profit (Windpower Monthly, 1998e).

Utilities can fulfil their renewables quota commitments in three ways: by developing their own renewable power plants, by negotiating bilateral agreements with independent producers, or by purchasing green labels on the open market. This mechanism is similar to the Renewables Portfolio Standard (RPS) mechanism being contemplated in the USA and essentially reserves a certain percentage of the electricity market for renewable energy within an otherwise liberalised market. However, unlike the RPS, the Dutch green labels scheme guarantees that all renewable generators can sell power to the grid at an assured price, thus removing some of the market uncertainty of the RPS but simultaneously perhaps reducing the economic incentive to reduce renewable energy costs.

The Dutch experience thus incorporates a wide variety of mechanisms for promoting renewables. Some current efforts such as the green funds and green labels programmes represent creative and exciting new initiatives. However, the Dutch programme has suffered some similar drawbacks as the US programme, such as lack of policy stability and difficulty for independent producers negotiating acceptable contracts with utilities. The past few years have seen Dutch wind energy activity drop off drastically as previous subsidies were eliminated and not replaced by sufficient new incentives. Initial activity in the green label market has provided generous prices (Windpower Monthly, 1998e), but whether such prices will be sustained and will provide the necessary financial stability to stimulate significant new projects remains to be seen. As of late 1998, few trades in green labels have actually taken place (Wolsink, 1998).

The Netherlands has also suffered similar wind turbine siting difficulties as in the UK, with local groups often opposing new developments in spite of a general philosophical support of green energy. This difficulty may be attributed in part to an emphasis on centralised utility-based wind energy development, which is not well-equipped for addressing many local planning concerns (Wolsink, 1996). Decentralised solitary wind turbines have also been the subject of protests, however, with the accusation that too many dispersed turbines lead to 'horizon pollution' (Windpower Monthly, 1998f). More recently, wind energy planning and implementation responsibilities have been shifting increasingly from central authorities to a more local level, and this is expected to facilitate wind plant siting to some degree (Wolsink, 1998).

With the combination of policy instability, declining incentives and siting difficulties, actual wind energy installation in the Netherlands has been well below official government targets in the past few years. Recently, the government reduced its target for installed wind capacity from 1000 MW down to 750 MW in the year 2000, but this revised target may also be difficult to achieve in practice.

## Denmark

Denmark's renewable energy promotion strategy has concentrated primarily on wind energy and to a lesser degree on biomass-based combined heat and power. Denmark's wind energy policies have been markedly different from those described above for the USA, UK and Netherlands. Stable policy has been a notable feature of Denmark's wind programme, providing a reliable home market unlike most other countries. This, combined with an emphasis on technology reliability, has enabled Denmark to become the world's dominant wind turbine manufacturing country, achieving a 60 per cent world market share in 1996 (BTM Consult and Danish Wind Turbine Manufacturers Association, 1996).

Incentives for wind energy in Denmark vary according to ownership, which can be divided into three categories: wind energy co-operatives, private ownership, and utility ownership. For co-operatives and private owners, incentives include the following: (1) guaranteed power purchase contracts with utilities in which utilities pay generators 85 per cent of the local retail price of electricity, amounting to approximately 0.33 DKK/kWh; (2) refund of 0.17 DKK/kWh energy tax; (3) refund of 0.10 DKK/kWh $CO_2$ tax. As a result, non-utility-generated wind power receives a total payment of approximately 0.60 DKK/kWh (Morthorst, 1996), or 0.091 US$/kWh at the average 1997 exchange rate of 6.608 DKK per US$.

Furthermore, individual persons who participate in wind energy co-operatives can own up to 20 000 kWh/yr -worth of shares in the co-operatives, of which the first 3000 DKK/yr of income is tax-free (and the remainder taxed at a 60 per cent rate). To the extent that the wind power purchase contracts increase the cost of electricity, these costs are passed on to utility ratepayers. Lastly, any grid reinforcement which may be required as a result of non-utility wind power installations is paid for by the utilities.

Utility-owned wind power projects do not benefit from preferential tax treatment or from any refund of the energy tax, though utilities can obtain refunds of the $CO_2$ tax. Less incentive exists for utilities to build wind projects than for co-operatives and private owners. Nevertheless, utilities are committed to building more wind power as part of an agreement with the Danish government. In 1995, approximately 30 per cent of total installed wind-energy capacity was utility-owned (Vindmølleindustrien, 1997). Total installed wind-energy capacity in Denmark in late 1998 was over 1400 MW, providing 9 per cent of total Danish electricity production.

A major difference between Danish wind turbine ownership and that of other countries has been Denmark's emphasis on local project ownership by individuals and through wind energy co-operatives, with lesser emphasis on large wind farms. This approach, while more expensive, has made it possible for local populations to benefit economically from wind power and has thus successfully reduced the local opposition to wind power development that has plagued other countries. Nevertheless, opposition is growing as the number of installations increases, and siting is becoming increasingly difficult given Denmark's small geographic size. New developments are therefore shifting towards offshore wind farms by utilities, thus potentially shifting development away from traditional small-scale local ownership.

Denmark's other emphasis in renewable energy has been with biomass, pursuing increased use of combined heat and power (CHP) produced using primarily straw and wood chips. This strategy complements Denmark's traditional emphasis on district heat and on combined heat and power within its fossil fuel-based generation. The extensive use of combined heat and power also adds a degree of operational flexibility to the Danish electricity system, making it more conducive to intensive use of wind power, as outlined at the end of Chapter 2.

Payment to biomass-based generators is not fixed at 85 per cent of the residential tariff as for wind, but instead varies by time of day based on the utilities' avoided costs. Payment conditions differ between western and eastern Denmark, but in western Denmark, for example, the electricity purchase price paid to biomass CHP-based generators differs between three set time-of-day periods: peak, mid-peak and off-peak. On average, payments to biomass-based generators are on the order of 10–20 percent lower than payments to wind generators, reflecting a higher subsidy for wind. The payments for biomass generation are, however, likely still to be higher than the utilities' actual avoided costs. As with wind, biomass CHP generators also receive refunds of the 0.17 DKK/kWh energy tax and 0.10 DKK/kWh $CO_2$ tax.

## Germany

Germany's promotion of renewable energy has concentrated primarily on wind and solar energy. In 1997 Germany surpassed the USA as the country with the largest installed wind energy capacity in the world. Germany's installed wind capacity has grown very rapidly, from negligible in 1990 to almost 2900 MW in late 1998. Three primary components have been responsible for this growth.

First and foremost is the Renewable Energy Feed-In Tariff (REFIT) contained within Germany's Electricity Feed Law (EFL). The REFIT specifies the price at which German utilities must purchase all power from renewable generators; and this price is tied to the residential electricity tariff. Wind generators receive a payment of

90 per cent of the residential tariff, amounting to a payment of 0.1721 DM/kWh in 1996 (Hoppe-Kilpper et al., 1996). At the average 1997 exchange rate of 1.735 DM per US$, this would be equivalent to 0.099 US$/kWh, approximately 10 per cent higher than the payment for wind provided in Denmark. The extra costs of purchasing this wind power compared to conventional electricity are passed on to electricity customers of the local purchasing utility, causing higher electricity prices in areas with substantial wind energy development. This is changing, however, to uniform funding by consumers throughout the country to reduce regional funding inequities.

Another major stimulus to wind energy development has been the '250 MW Wind Programme', which was started in 1990 as a large-scale demonstration programme which would pay developers for their output or for approved investment costs. The programme provides investment subsidies of DM 200/kW with a ceiling of DM 100 000 for each project (DM 150 000 for projects with facilities greater than 1 MW) (Lindley, 1996). For a 600 kW wind turbine costing on the order of DM 1700/kW, the maximum subsidy would amount to approximately 10 per cent of capital costs.

The third component of Germany's wind promotion programme comes in the form of preferential financing. Below-market loans are available from the Deutsche Ausgleichsbank (DtA), a federal funding institution which provides favourable financing terms for projects in areas such as environmental protection. In conjunction with the European Recovery Programme (ERP) Fund, DtA loans are available at a fixed interest rate of 1–2 per cent below commercial rates; and a maximum repayment grace period of five years is allowed to ease cash-flow constraints during projects' initial years. With such DtA/ERP loans covering approximately 75 per cent of total project cost, combined with another 12–15 per cent of project cost funded by local bank loans, and approximately 5 per cent of costs covered by grants, investor equity requirements are limited to a mere 5 to 8 per cent of project cost (Lindley, 1996). This contrasts with the at least 20 per cent and even 50 per cent equity fractions seen in project-financed projects in the USA (Kahn, 1995).

The guaranteed power purchase contracts, generous per-kWh payments and highly favourable financing terms combine to make Germany a very attractive market for wind project developers and explain the programme's significant success in stimulating wind energy utilisation. However, some observers have criticised the programme as being overly generous; and utilities in particular have been opposed to the high price they are forced to pay for wind energy regardless of their actual need for the energy. Such concerns are becoming particularly prominent in the light of moves towards electricity industry liberalisation throughout Europe, which could place German utilities at a competitive disadvantage against other European utilities. The German parliament therefore came close to amending the REFIT in 1997 to drastically reduce power purchase prices for wind projects, though this did not actually come to pass.

Thus, Germany's wind energy programme has exhibited a positive combination of stable policy during the 1990s, generous per-kWh payments to encourage maximum energy production, easy financing conditions to overcome commercial finance institutions' risk-aversion, and modest capital subsidies which help stimulate investment but are not high enough to invite abuse. In this regard, the great success of Germany's programme is easily understandable. However, whether Germany has received good value for money is less clear. Some would argue that Germany's support has been too generous and that the German state is subsidising large, virtually risk-free, profits for private investors. It is almost certain that the level of German subsidies has been higher than what could be justified on the basis of avoided costs or environmental benefits, though long-term 'strategic' benefits are harder to quantify. Notably lacking in Germany's programme has been any pressure for project developers to reduce costs. Instituting some form of competition or gradually ratcheting down premium payments could greatly improve the cost-effectiveness of German wind projects without necessarily damaging the industry.

Though wind has been Germany's greatest renewable energy success, Germany's support for renewables is not limited to just wind. Solar energy has been the other large thrust of the German renewables programme, receiving two-thirds of German renewable energy R&D funding (while wind receives most of the remaining one-third). In particular, photovoltaic programmes have received strong emphasis. In terms of power purchase prices, solar energy, like wind energy, receives a payment of 90 per cent of the average residential electricity tariff (other sources, like biomass and hydro, receive less). Solar energy installations also receive further direct support from both the German Federal and state governments.

## India

India has been supporting renewable energy development since the late 1980s through the Ministry of Non-Conventional Energy Sources (MNES) and interested state governments. In terms of grid-connected renewable electricity, India's efforts have also been focused primarily on wind, though biomass-based cogeneration has begun receiving significant attention more recently.

In late 1998, India had approximately 1000 MW of installed wind capacity and ranked third worldwide in wind power installations after Germany and the USA. Early wind power development was largely accomplished through demonstration projects by the MNES, but this has given way to private development which now accounts for the vast bulk of installed capacity. Indian wind energy policy has been successful not only in achieving significant capacity installations but also in stimulating the development of a domestic wind-turbine manufacturing industry. The longer-term prognosis for future development is less certain, however, due partly to Indian climatic conditions and the prevalence of winds primarily during the low-demand rainy season.

India's support for wind energy is characterised by the following incentive mechanisms: guaranteed power purchase arrangements, tax incentives and concessional loans, though actual rules vary by state (Sarkar and Bhatia, 1997). Power purchase arrangements can typically be handled in any of three ways. The developer/generator can use its generated wind power by wheeling it through the utility grid to its own industrial facilities (and paying a wheeling charge), the generator can sell the power to the state utility, or the generator can sell the power to a third party, again paying a wheeling charge. Because India's winds are seasonal and largely occur during the monsoon season, banking options are also available, in which the generator can bank or 'store' the wind power by distributing it to the grid at the time of generation and claiming it back later in the year as the need arises. Standardised power purchase rules exist in each state, eliminating the need for more complex individual negotiation of contracts. Conditions vary by state, but, as an example, the following summarises the rules in effect around 1997–98 (Sarkar and Bhatia, 1997) for two states, Tamil Nadu and Andhra Pradesh, which have been among the most active in promoting wind energy in India.

- *Tamil Nadu.* Electricity could be wheeled through the utility grid for a fee of 2 per cent of energy generated. Electricity banking was allowed for 12 months at 2 per cent banking charges. The electricity buy-back price was 2.60 Rs/kWh in 1998, with 5 per cent annual escalation until the year 2000 (as of November 1998, 1 US$ = approximately 43 Rs). Wind generators were also exempt from the state's electricity generation tax.

- *Andhra Pradesh.* Electricity could be wheeled through the utility grid for a fee of 2 per cent of energy fed to the grid. Electricity banking was allowed for 8 months between August and March. For 12 months of banking, 2 per cent charges apply. The electricity buy-back price was 2.60 Rs/kWh in 1998. Sale of the wind-generated electricity to third parties was allowed. Capital subsidies were available for 20 per cent of project cost up to a maximum of Rs 2.5 million. Land could be leased for 20 years, and was free of rent for the first five years.

However, by 1998 the state electricity boards (SEBs) of Tamil Nadu and Gujarat states (which together account for more than 90 per cent of installed wind power capacity in India) had stopped allowing independent power producers to wheel electricity through their grids directly to customers. This was done because the SEBs were losing many of their best customers to third-party sales and were not recovering this loss of revenue through their modest wheeling charges (Windpower Monthly, 1998g). With many SEBs throughout India on the verge of bankruptcy and unable to guarantee long-term power purchases, the elimination of wheeling provisions could prove to be a very serious threat for continued wind power development.

In addition to incentives from state governments, further tax incentives are available from the central government. These take the form of 100 per cent accelerated depreciation in the first year of wind farm commissioning, as well as duty-free or reduced-duty import of wind turbine components. These tax concessions have proven to be highly effective incentives, though recent reductions in corporate tax rates, from 46 per cent in 1994 to 30 per cent in 1998 (Windpower Monthly, 1998g), have significantly lessened their impact. In addition, such capital-cost incentives did incur some initial abuses through use of less effective second-hand turbines, but such abuses have been eliminated through tightening of eligibility and commissioning rules. More recently, however, further scandals have come to light involving allegations of tax evasion in which companies are accused of having falsified records to claim the 100 per cent first-year depreciation without actually installing any wind turbines (Windpower Monthly, 1998e). Such allegations serve as another reminder that capital cost-based incentives and tax credits are vulnerable to abuse and require very careful monitoring by a vigilant regulatory authority.

India has also established the Indian Renewable Energy Development Agency (IREDA), a public limited government company under the MNES, specifically for financing renewable energy projects. IREDA wind energy loans are available for 100 per cent of eligible equipment cost, limited to a maximum of 75 per cent of total project cost. Loan terms have been for ten years, with a repayment grace period of one year (IREDA, 1997). Their interest rates of 15–16 per cent were considered concessional in earlier years when commercial rates were higher; but in recent years commercial interest rates have declined and have thus made IREDA loans less attractive (Suresh, 1997). As a result, IREDA interest rates have more recently been reduced by 1 per cent while the repayment period has been increased (Suresh, 1999).

IREDA provides financing for all forms of renewable energy, not just wind. For biomass cogeneration (the current focus of increased interest for independent power production), IREDA interest rates are similar to those for wind, ranging between 15.5 and 17 per cent, depending on the technology. Loan conditions are also similar to wind, with IREDA financing for biomass being limited to a maximum of 75 per cent of total project cost. Loan terms range between five and ten years, but repayment grace periods are provided for two to three years (IREDA, 1997).

In general, IREDA interest rates tend to be in the 15–17 per cent range, but lower interest rates are allowed for certain technologies and applications to which the Indian government attaches particular priority. Solar hot-water heaters can be financed at interest rates ranging from 2.5 per cent to 8.3 per cent, for example. Biogas plants based on animal and/or human waste can be financed for rates between 4 per cent and 10.5 per cent; and some rural photovoltaic and wind projects can be financed at between 2.5 per cent and 8.5 per cent (IREDA, 1997).

## Sweden

Swedish electricity policy has been in a significant state of flux over the past several years as the industry has been liberalised and opened up to competition throughout the Scandinavian Nord Pool market. Renewable energy policy has thus also been changing. Between 1991 and 1996, Sweden provided capital-cost subsidies for the following technologies (IEA, 1996a):

- *Wind Energy.* Subsidies of 35 per cent of capital costs for new wind turbines over 60 kW, with total available funding of SEK 350 million (as of November 1998, 1 US$ = approximately 8.2 SEK).

- *Solar Energy.* Subsidies of 25 per cent of costs of large-scale solar projects and technology development, with total available funding of SEK 136 million.

- *Biomass Energy.* Subsidies of 4000 SEK/kW for new biomass CHP plants, or 25 per cent of capital costs for conversions of existing facilities to use biomass fuels, with total available funding of SEK 1 billion.

Expiration of capital-cost subsidies in 1996 significantly slowed the implementation of many new renewable energy installations, but the Swedish government has again included further subsidies in its new energy law which took effect in February 1998. The new law provides subsidies as a percentage of capital cost as follows: (a) biomass-based cogeneration: 30 per cent; (b) wind: 15 per cent; (c) small-scale hydro: 15 per cent (CADDET, 1998d).

Other than capital-cost subsidies, two other primary mechanisms exist for supporting small renewable energy projects such as wind within the liberalised Swedish electricity market. The first is guaranteed power purchase contracts with local utilities. Prior to electricity market reform, utilities holding regional power concessions were required to purchase electricity at their avoided cost from all small power projects with generation capacities of up to 1500 kW. This requirement continues to exist under the new law, but the price now paid to small generators is equal to the residential tariff, plus a credit for reduced transmission and distribution losses, minus reasonable costs for utility administration and profit. In 1996 the average price paid for wind electricity was 0.26 – 0.28 SEK/kWh (IEA, 1996b). However, this power purchase requirement for small generators is limited in duration to five years, and subsequently all power producers are expected to compete on the open market. Whether small power producers will continue to survive at that time remains to be seen.

The other support mechanism for wind energy is an environmental bonus (SEK 0.138/kWh in 1997) paid from the government (IEA, 1996b). The amount of this bonus corresponds to the tax charged for household electricity consumption.

For biomass, in addition to the capital subsidies described earlier, other subsidies have included payments of 10–15 per cent of the costs of connecting small biofuel-based boilers in the industrial and residential sectors to district heating networks in 1994–95. Environmental taxes have been the other major force in encouraging increased use of biomass. Heating fuels are taxed for sulphur, $CO_2$, and $NO_x$ emissions, as well as being subject to a general energy excise tax. The energy excise tax (for example, 251 SEK/ton coal), $CO_2$ tax (367 SEK/ton $CO_2$ [approximately 48 US$/ton $CO_2$], or 916 SEK/ton coal) and the sulphur tax (for example, 30 SEK/kg sulphur for coal and peat, or 150 SEK/ton coal) add substantially to the cost of burning fossil fuels (IEA, 1996a). Biomass fuels are exempt from the energy, $CO_2$, and sulphur taxes, though peat does incur the sulphur tax.

Note, however, that the above-mentioned taxes are for heating fuels. Fuels used in electricity generation are not subject to the energy excise tax nor $CO_2$ tax, though they are subject to the sulphur tax. Because nuclear and hydro power provide almost 95 per cent of Sweden's electricity, the exemption of fossil fuels for electricity from environmental taxes has not been of major importance. However, this exemption has led to strange practices in biomass-based CHP plants. Given that fossil fuels are subject to environmental taxes for heating but not for electricity, biomass-based CHP plants have burned both fossil fuels and biomass within the same plant, claiming that the biomass is used for the heating portion and the fossil fuels for the electricity portion. Other distortions also exist. For example, with industrial facilities being exempt from a significant portion of the $CO_2$ tax (for reasons of international competitiveness), many industries have been shifting away from burning biomass fuels, choosing instead to burn fossil fuels and sell their biomass to district heating plants who are subject to the $CO_2$ tax. This highlights some of the complexities of using the tax system to achieve public policy goals. Further harmonisation of taxation laws may therefore be necessary.

In addition to the above direct and indirect support mechanisms for renewables in the marketplace, the Swedish government also funds energy technology research programmes as well as a development and demonstration programme which provides support of up to 50 per cent of cost for demonstration projects including solar and wind plants, for example. The National Board for Industrial and Technical Development (NUTEK) has also organised a technology procurement programme to try to further reduce the cost of wind power by forming a consortium of purchasers, clearly specifying technical and economic requirements, developing a financing scheme with a bank consortium, and guaranteeing a minimum purchase order to the turbine manufacturer who wins the bidding competition. This process is based on the successful technology procurement concept practised by NUTEK in promoting energy efficiency.

# 8
# Summary and Conclusions

Installed wind power capacity has been increasing at an average rate of over 25 per cent per year between 1992 and 1997, making wind energy the world's fastest growing energy sector. This growth rate shows few signs of slowing down. On the contrary, installed wind capacity has grown by well over 30 per cent per year between 1998 and 2000, surpassing 10 000 MW in 1998 and 18 000 MW in 2000. Installed capacity is expected to continue growing on the order of 20 per cent per year until 2007, with total worldwide wind capacity expected to approach 50 000 MW by then (BTM Consult, 1998a). This anticipated growth rate is of a similar order to that achieved by nuclear power between 1968 and 1977, during which time installed nuclear capacity increased from 9200 MW (similar to installed wind capacity in late 1998) to 99 000 MW, an annual rate of increase of 30 per cent (Worldwatch, 1999).

This growth is expected to be spread around the world, with major roles played by Europe, the USA, India and China. The USA accounted for most wind energy installations in the 1980s, but the industry's centre of gravity shifted to Europe in the 1990s, both in terms of installations as well as manufacturers. The European Union's renewable energy strategy aims to increase the EU's installed wind plant from 4500 MW in 1997 to 40 000 MW by 2010 (European Commission, 1997). Developing countries, led by India and China, have also achieved significant wind energy growth in the late 1990s. And after nearly a decade of stagnation, installations in the USA began increasing significantly again in 1998.

Looking further ahead to 2020, the World Energy Council anticipates wind energy installations of between 180 000 and 480 000 MW (WEC, 1994), depending on the demonstrated level of ecological concern and commitment to climate-change mitigation. Achieving even a fraction of these installations would signify the emergence of a multibillion-dollar wind energy industry of truly global proportions. Global wind turbine sales in 1998 were already valued at around US$2 billion.

Wind turbines come in horizontal- and vertical-axis configurations. During the 1990s commercial wind turbine designs settled into a 'standard' fundamental concept of horizontal-axis machines with two or three blades (three being much more common in large-scale turbines), rotating at near-fixed speed. Great advances were achieved in efficiency and reliability, as well as in economies of scale, both in terms of increased turbine size and increased manufacturing volume. Turbine sizes, for example, have increased by a factor of ten over the past decade, from 150 kW in 1989 to 1.5 MW and higher in 1999. All of these improvements have led to drastic reductions in wind energy's cost per generated kilowatt-hour. These advances to date have been essentially evolutionary in nature, based on perfecting a relatively constant basic design.

The currently predominant basic design is by no means the 'ultimate' wind turbine design, however. Significant advances remain to be made in making wind turbines cheaper, lighter, more flexible and more efficient. Many of these advances are becoming possible through increased computing power and improved understanding of wind turbines' aeroelastic behaviour. In terms of drive trains, advances such as variable-speed drives have already been introduced and are expected to continue. Improved and cheaper power electronics should also increase wind turbines' overall flexibility and efficiency in operating under a wide variety of conditions.

These technological advances will not happen overnight, but will more likely evolve over time. On the other hand, the wind industry's explosive growth is beginning to entice new deep-pocketed corporations into the market, and these new entrants may have a stronger incentive to introduce radical new designs in an effort to gain market share from older established players. As a result, the next two decades are expected to witness vigorous competition amongst manufacturers and designs.

Reductions in the per-kWh cost of wind-generated electricity are due to a combination of factors, including lower capital costs, lower operation and maintenance costs, higher generation efficiency, higher reliability and improved siting. Overall, from the late 1980s to the late 1990s, wind power costs per kWh have decreased by approximately 45 per cent in less than a decade. Using today's advanced turbines, wind power costs for large-scale grid-connected turbines are typically in the range of 4–5 US cents/kWh, and sometimes even in the 3–4 cent/kWh-range under favourable wind conditions. As a result, wind energy is becoming close to competitive against conventional electricity sources such as coal, nuclear and natural gas, and in some cases wind energy is already cheaper than conventional sources on a total-cost basis.

Regarding offshore wind turbines, their costs are still higher than those on land but are expected to decline significantly as some European countries aggressively pursue this option and gain further experience. Denmark, for example, plans to build more than 4000 MW of offshore wind power plants by the year 2030.

In spite of their improved cost-effectiveness, wind power plants often have much greater difficulty getting built than do conventional power plants, in part a result of lack of adequate financing. This is due to several factors, but one of the most significant is investors' perception of higher financial risk associated with wind power. This higher perceived risk results in not only a higher financing cost (raising wind power's overall cost), but in some cases a lack of available finance altogether. Some perceptions of risk stem from memories of certain wind turbines' poor technical performance during the 1980s and are no longer justified in light of the high reliability of current wind technology.

Other risks are real, however, including the inherent variability of the wind itself. As an intermittent resource, wind power cannot be simply turned on and off according to need; and a wind turbine cannot necessarily be assumed to operate during times of high electricity demand. As a result, not only is a unit of generation capacity from a wind turbine inherently less valuable than an equivalent unit of capacity from a dispatchable resource like a gas turbine, but wind plants also have much greater difficulty obtaining power purchase contracts than do conventional power plants. Availability of reliable power purchase agreements is therefore a key consideration in terms of wind power's financial viability as well as overall wind energy policy.

The increasing trend towards competitive generation markets introduces further special considerations. In countries that incorporate independent power producers (IPPs) into the generation market while maintaining a utility monopoly in transmission, distribution and wholesale/retail sales, the opening of the generation market to private players is likely to provide increased opportunities for wind power. This trend towards greater reliance on IPPs is notable in many developing countries around the world.

In systems of wholesale and retail competition increasingly favoured by developed countries, generation markets typically consist of two coexisting contract types: long-term bilateral contracts between buyers and sellers, and short-term auction-type forward and spot markets. Short-term forward markets require generators to submit bids to sell power in advance, typically one day ahead of the sale. Wind plants' ability to sell in the short-term markets is therefore critically dependent on their ability to accurately predict wind speeds and wind turbine power output in advance.

Fortunately, advances in wind energy have not been solely in the area of turbine technology. Significant improvements have also occurred with wind prediction, and wind speeds can now be predicted 24 to 36 hours in advance with an accuracy of around +/– 20 per cent, using sophisticated current techniques. This level of accuracy allows wind plants to bid their power into day-ahead forward markets, but a wind plant's bid will nevertheless be much less accurate than a bid from a conventional generator.

As a result, the other critical factor for wind turbines to function in a short-term forward market is the market's rules regarding penalties for generators who over- or under-generate compared to their bid. Markets which impose severe penalties on generators are not only likely to be less efficient overall, but they will severely limit the ability of intermittent generators like wind to compete. A more efficient market system includes a separate market for 'balancing' power to make up for any instantaneous differences between demand and supplied power. An example of this is found in the 'regulation market' of the Nord Pool system in Scandinavia. Such market systems allow intermittent generators such as wind turbines to sell to the short-term market with only minor cost penalties for their inherent variability.

Wind plants also face greater difficulties obtaining long-term bilateral contracts due to their intermittent nature. Bilateral contracts are typically signed to reduce risk by locking in the sale/-purchase of a given amount of power for a set price and time. Because wind plants cannot guarantee their power output at any given time in advance, purchasers have fewer incentives to sign bilateral contracts with wind generators.

On the other hand, through the Kyoto Protocol to the United Nations Framework Convention on Climate Change, industrialised countries have committed themselves to reductions in their emissions of greenhouse gases such as $CO_2$. Binding commitments by countries to reduce $CO_2$ emissions must surely lead to some form of market for $CO_2$ emission reduction credits. Once such markets begin to develop, power purchasers may begin signing bilateral power contracts with wind generators – not to lock in a fixed quantity of power, but rather to lock in $CO_2$ emission credits.

Thus, while the move towards competitive generation markets may create significant complexities for wind power, diverse factors such as improved wind prediction and $CO_2$ emission reduction commitments may result in wind power plants being able to successfully compete for both short-term forward market sales as well as long-term bilateral contracts.

The discussion of $CO_2$ credits brings up the issue of the environment in general. The environment is clearly the ultimate driving force behind the development and implementation of wind energy. Yet, conventional electricity markets place only limited emphasis on environmental considerations. The fact that electricity prices do not reflect the environmental damages caused by power generation is a major deterrent to increased adoption of wind energy.

The valuation of environmental externalities is a controversial topic and subject to great uncertainty. The uncertainty and controversy are particularly great regarding potential damage costs of global climate change, the area of wind energy's greatest environmental advantage over fossil fuels. Nevertheless, major studies in both Europe and the USA suggest that the environmental damages avoided through wind energy could be worth several US cents per kWh. Wind energy would clearly be competitive against fossil fuel-based generation if this cost of environmental damages were reflected in electricity prices. It is hoped that implementation of the UN Framework Convention on Climate Change will lead to some form of $CO_2$ credit market which will allow wind energy to reap the financial benefits of its environmental advantage.

In spite of wind energy's environmentally benign nature, wind energy does also create a number of potential environmental impacts, including visual intrusion, noise and bird deaths. Noise and bird deaths are being addressed through improved design and turbine siting, and these are not expected to cause significant overall impediments to wind energy development. Visual intrusion, on the other hand, has received significant public attention and has been a major stumbling block for wind energy development in some locations. However, repeated studies indicate that the actual visual harm caused by wind turbines is for the most part quite small, and that opposition to wind development is often led by only a small but vocal minority. Nevertheless, aesthetic concerns and related local planning considerations are important issues which must be addressed with sensitivity by wind power developers.

The advent of electricity industry restructuring and retail competition is bringing about another interesting development in the area of environmental protection. This is the concept of green marketing, in which consumers choose to purchase electricity from renewable energy generators such as wind, often voluntarily paying a higher price for this service. This market-based approach to environmental protection is subject to some controversy, and it is still too early to ascertain how much renewable energy development will actually occur as a result of such programmes. However, green marketing programmes have proven to be popular in the Netherlands, for example; and green marketing appears to be developing into a potent competitive force in the residential sector under the USA's ongoing electricity-industry restructuring process.

Many of the advances in wind energy over the past two decades have been made possible through changes in the policy environment and the application of innovative new incentive mechanisms. These mechanisms include guaranteed power purchase agreements, investment incentives, production incentives, market set-asides, externality adders, environmental taxation, R&D grants, government-assisted business development, green marketing, wheeling provisions and electricity banking. Different countries have had varying degrees of success with such mechanisms, and it is instructive to examine the experiences of countries who have pursued wind energy development.

The USA has in the past relied primarily on tax incentives, and guaranteed power purchase contracts through the Public Utility Regulatory Policies Act, while newer initiatives include green marketing and market set-asides through a proposed Renewables Portfolio Standard. Though US policy initiatives were largely responsible for the creation of the modern wind energy industry in the 1980s, US renewable energy strategy has suffered from a lack of policy stability and the resulting high uncertainty and risk for developers and investors. The Netherlands has also pursued a variety of innovative incentive mechanisms such as green investment funds, tax incentives, green marketing, and creation of a market set-aside using tradable 'green labels'. Nevertheless, the Netherlands has also suffered from unstable policy as well as turbine siting difficulties, resulting in mixed success in stimulating wind energy development.

Denmark and Germany are two of the world leaders in wind energy, and both of their strategies have been characterised by stability and simplicity. In both countries, the basic incentive structure consists of guaranteed power purchases of all wind-generated electricity at a set (and generous) price. Other incentive structures in place include $CO_2$ taxation in Denmark and a preferential finance facility in Germany. While these countries have had great success in encouraging wind energy development, it is unclear whether their incentive systems have provided good value for money; and some observers argue that wind generators have obtained excessive profits at the expense of electricity ratepayers.

The UK has pursued an innovative, competitive, market-based incentive system known as the Non Fossil Fuel Obligation (NFFO). This is the best known of the market set-aside systems and has been highly successful in stimulating wind energy development as well as encouraging cost reductions, while managing to coexist with a competitive deregulated generation market. The UK has faced significant public controversy regarding the local visual impact of wind turbines on the environment, however, and some observers attribute this controversy in part to the highly price-competitive nature of the NFFO process and the resulting concentrated development in windy but scenic locations.

Other countries reviewed include India and Sweden. The key component to all countries' success in stimulating wind energy development has been the availability of stable and reliable power purchase contracts. This is a particularly important issue which must be carefully considered by countries attempting to increase competition in their electricity generation markets.

Overall, wind energy represents a great success story in which, over a brief span of time and perhaps alone among renewable energy technologies, wind power is on the verge of making the transition from alternative energy source to an integral part of the mainstream electricity industry. However, this transition is not yet complete, and the next five to ten years will represent a critical time in the wind industry's maturation. At this time, wind energy continues to require a favourable policy environment to encourage additional implementation, continued technological development and further cost reductions. A key element of this will include more explicit recognition of wind energy's superior environmental attributes and a means by which wind power plants can be compensated for this benefit. To help complete wind energy's successful emergence as a mainstream electricity source, this book recommends that policy makers undertake the following actions:

1. **Provide stable markets for wind-generated electricity.** Reliable markets for wind-generated electricity are the single most important factor for stimulating the further development of wind energy. Stable power purchase contracts have been a critical feature of the energy policies of all countries who have achieved wind energy success. The need for stable contracts is amplified by the emergence of competitive generation markets and the increased risk which such markets entail. In general, it is critical that policy makers understand the risks facing wind energy developers and create a policy environment which helps manage these risks. A variety of mechanisms exist to help achieve a degree of contract stability within the context of a competitive generation market. Possibilities include an auction-based set-aside market such as the UK's Non Fossil Fuel Obligation, or a set-aside market based on tradable renewable energy credits, such as the Netherlands' Green Labels programme or the USA's Renewables Portfolio Standard. Incentive mechanisms which reward efficient actual production of electricity are likely to be more effective than incentives based on capital investment.

2. **Provide stable wind turbine markets.** As a corollary to the above recommendation, countries should not only aim to

provide stable power purchase contracts but should also encourage a degree of stability in markets for new wind turbines to ensure the industry's viability and to stimulate technological development. It is not necessary for this market to be large, as market reliability is more important than market size. Even a limited (but stable) turbine market can successfully encourage technological development, as long as it provides appropriate performance incentives. The boom-and-bust cycles which have characterised turbine markets to date have placed great strain on the industry and have hampered long-term development.

3. **Align energy projects' financial performance with society's environmental goals.** Traditional energy policies have not taken full account of the environmental benefits associated with wind energy, thus favouring more highly polluting energy sources. Efforts should be redoubled to introduce appropriate mechanisms which will better align energy projects' financial performance with society's environmental goals. Pollution taxes, for example, are an effective mechanism for this. They not only benefit the environment but also encourage more efficient economic development by discouraging wasteful practices, stimulating technological development, and reducing the need to tax other more beneficial and income-producing activities. Countries should also move forward with coherent strategies for addressing climate change, which should help create a substantial market for $CO_2$-free energy such as wind.

4. **Enhance community participation in project planning and in reaping project benefits.** Visual and noise impacts of wind energy can cause local objections and make wind turbine siting difficult. These visual and noise impacts, while generally quite low, are nevertheless real and must be addressed through an open and straightforward planning process. Improved information and greater familiarity with wind projects goes a long way towards reducing local opposition. Opposition is also considerably reduced when local residents are able to benefit financially from wind energy projects in their communities. Enhanced community participation should be a goal of all public infrastructure projects, and wind energy is no exception.

5. **Encourage decentralised projects in remote communities.** This book has concentrated primarily on large-scale grid-connected electricity, where the most dramatic advances have taken place. However, wind energy can be highly cost-effective in remote communities not served by a central electricity transmission grid. Because such communities are often poor and in developing countries, less effort has been made to address these markets. Additional efforts should be directed towards making financing available and stimulating off-grid wind energy projects in remote communities, particularly in the context of rural electrification programmes in developing countries.

6. **Remove institutional barriers to wind energy.** Institutional barriers can be as significant an impediment to wind energy as technological or economic barriers, but are often left unaddressed. The first and foremost priority in removing such barriers should be the development of stable professional communities which understand wind energy issues and can facilitate their countries' long-term wind energy development. Development of institutional capacity also includes information dissemination, development of appropriate planning processes, quality certification programmes, and perhaps wind energy demonstration programmes where no projects yet exist.

7. **Encourage research and development, particularly for wind resource assessment.** Research and development funding for wind energy should be targeted towards those areas in which private investment is not readily available. Much technology development is already successfully addressed by the private sector and may not require significant public funding. Rather, accurate wind mapping and improved understanding of countries' wind resource potential would greatly help to enhance countries' understanding of the level of feasible wind energy development and identify suitable sites for private investment.

# Epilogue

Since the completion of the writing of this book, there have been a number of developments which merit an update. First, the cost-effectiveness of wind energy has continued to improve. In March 2001, it was reported that a 300 MW wind farm development in Oregon, USA, had achieved total costs of under $0.025 per kWh[1] (AWEA, 2001), further confirming wind energy's status as a viable competitor to conventional fossil-fuelled generation.

Perhaps the most striking development, however, has been the continued evolution of the California electricity market and its implications for markets worldwide. Chapter 7 describes the California market as of 1999 and the advent of retail competition. From the commencement of the competitive market in March 1998 until mid-2000, the new California market functioned more-or-less as designed, with vigorous competition in the wholesale generation market, wholesale prices averaging between 0.02 and 0.03 US$/kWh, and similarly vigorous competition to attract customers at the retail level.

From mid-2000, however, serious flaws in the market's design became increasingly apparent. A combination of factors, including growing electricity demand, lack of new power plant construction, low hydroelectric power availability, skyrocketing natural gas prices, lack of adequate price signals to consumers, and alleged market manipulation by generators and power marketers, led to electricity shortages and dramatically increased wholesale power prices. For example, whereas the average unconstrained market clearing price in the California Power Exchange day-ahead market was 0.03 US$/kWh during December 1999, by December 2000 the same average day-ahead price had risen to 0.38 US$/kWh, a more than twelve-fold increase.

California's 1996 electricity industry restructuring law committed its utilities to purchasing their power in the volatile short-term forward and real-time spot markets but imposed a rate freeze which prevented the utilities from passing rising wholesale power costs onto retail consumers. As wholesale costs began to far outstrip retail prices, California's major investor-owned utilities plunged into debt. And with the impossibility of making money under this structure, virtually all competitive non-utility retail electricity service providers abandoned the market, returning customers to the default utilities (thereby forcing ever greater losses on the utilities) and effectively ending the much vaunted competitive retail market in California.

By early 2001, the state's two largest utilities had collectively lost more than US$10 billion over the course of seven or eight months and were on the verge of bankruptcy. In spite of these high wholesale prices, power supplies were frequently insufficient to meet demand, and the state began to suffer periodic state-wide rolling blackouts. The utilities, unable to pay their bills, then began refusing to pay generators, who had to be forced by the state and federal government to continue to supply electricity to California. Only dramatic intervention by the government prevented a complete collapse of the entire California electric system. In April 2001, the state's largest utility, Pacific Gas & Electric Company, filed for bankruptcy protection from its creditors, thereby initiating the largest-ever utility bankruptcy in American history.[2]

California's electricity débâcle is likely to cost the state tens of billion of dollars in higher electricity costs, not to mention the additional billions in lost economic output due to power outages and resultant business uncertainties. The crisis has emerged as perhaps the single greatest threat to the prosperity of one of the world's most dynamic economic zones, and its economic impact will surely be felt for many years to come.

But what do these developments mean for wind energy? On the positive side, wind energy emerged as perhaps the lowest cost electricity resource among all fuels and technologies. While wholesale natural gas prices had averaged approximately 2 US$/mmBTU (million British thermal units) throughout most of the 1990s, the average price doubled to approximately $4/mmBTU during 2000. By December 2000, wholesale prices surpassed $10/mmBTU, a previously unimaginable level; and during one day's panic buying, prices hit $69/mmBTU at the Southern California border (California Energy Markets, 2000). At $10/mmBTU, even the most efficient natural gas-fired generation plant would have a short-run operating cost (not including capital costs or ongoing maintenance costs) of

$0.06 per kWh, higher than the total all-inclusive generation cost of a wind power facility.

This served to highlight not just wind power's overall cost-effectiveness, but also its diversity value in reducing the electricity system's over-reliance on one fuel. After many years of gas price stability, the world was reminded once again that all fossil fuels can be subject to price shocks of far-reaching impact. Whereas natural gas's dominance as the fuel of choice for electricity generation had come to be seen as inevitable during the 1990s, this rationale has suddenly been called into question.

Furthermore, some of the factors underlying the crisis are by no means unique to California; similar concerns are emerging in markets as diverse as New York and Brazil. The world's movement toward unfettered competition in electricity markets is thus beginning to be questioned, thereby reinvigorating the role of longer-term energy policy. In California, this policy direction is, for now, unambiguously pointing towards greener energy such as wind. In trying to resolve the electricity crisis, the California state government authorised the spending of close to $1 billion in March–April 2001 in new spending on energy efficiency and renewable energy, including low interest loans for wind and other renewable energy, and capital cost subsidies of up to 50 per cent of project cost (California Assembly Bill No. 29, 2001; California Senate Bill No. 5, 2001; CPUC, 2001). In many ways, the economic climate for wind energy has never been better.

On the negative side, wind generators and other smaller-scale PURPA Qualifying Facility (QF) generators were severely hurt by the California utilities' refusal and/or inability to pay for their electricity purchases as the market spun out of control, forcing many smaller generators to shut down operations entirely. Small QF generators were further damaged by the California Public Utilities Commission's arbitrary and politically motivated recalculation of electricity prices payable to QFs at below their short-run operating cost (FERC, 2001, California Energy Markets, 2001).

While natural gas-based QF cogenerators were particularly damaged by these events, wind generators were similarly harmed during a time when large wholesale generators were earning unheard-of-profits. In spite of wind generators' role in providing some of the lowest cost power available in the state during critical electricity shortages, their lack of a strong voice in a highly politicised regulatory environment meant that they were much less able to benefit from high wholesale electricity prices than were the larger fossil-fuelled generators.

And so once again in the USA, wind power continues to be buffeted by an uncertain regulatory and political environment. Market and policy stability, whose critical importance was underscored in Chapters 7 and 8, remain as elusive as ever. Yet, in spite of this, wind power continues to advance both technically and economically. In Europe, the USA, and elsewhere, installed wind capacity continues to increase at its breathtaking pace, and wind's place within the mainstream energy industry is becoming ever more secure with each passing year. All in all, the future of wind energy is brighter than ever.

# Notes

## Notes to Chapter 2

1. Electric generation capacity is presented throughout this book in terms of kW, MW, GW and TW. For readers not familiar with this terminology, 1 kilowatt (kW) = $10^3$ watts, 1 MW = $10^6$ watts, 1 GW = $10^9$ watts and 1 TW = $10^{12}$ watts. 1 kWh represents 1 kW generated for 1 hour and similarly for MWh, GWh and TWh. A TW-yr represents one terawatt (TW) generated for one year.
2. 'Annual mean wind speed' is a standard term applied based on 'long-term average', as opposed to 10-minute or half-hourly averages.
3. Acknowledgement for this section: Lars Landberg, Niels Gylling Mortensen and Erik Lundtang Petersen, Wind Energy and Atmospheric Physics Department, Risø National Laboratory, Denmark.
4. For a complete description of the wind atlas methodology, see Troen and Petersen (1989).
5. The wind rose is a graphical representation of the relative frequency, average wind speed and energy content of the wind from each direction: north, north-east, east, south-east and so on. The wind rose is typically drawn with 12 sectors, each sector representing an arc of 30 degrees on the compass.
6. Some of these constraints are already being experienced in countries such as the UK, Denmark, Netherlands, USA and Germany. See Chapter 6 for further discussion of these environmental considerations.
7. This growth rate is slightly different from that shown in Table 2.1, due to accounting differences.
8. Capacity factor is defined as the total energy produced by a facility in a year divided by the total energy which could theoretically be produced by the facility if it operated at full rated capacity for the full year (see Swisher et al., 1997).
9. Syngas is the product of a gasification process, typically derived from coal, but also from biomass.

## Notes to Chapter 3

1. The stream tube is defined by the stream lines following the edge of the wake. Therefore, there is no flow perpendicular to the streamlines.
2. This section is based on Andersen and Jensen (1997).
3. A wind turbine's availability is defined as the percentage of time a wind turbine is capable of generating electricity without manual intervention.

4. Availability was defined in note 3.
5. See note 8 in Chapter 2.

## Notes to Chapter 4

1. Note: economic and financial figures are typically presented in US dollars. Unless specifically noted otherwise, currency conversions in this book have been made using the following average 1997 rates: US$1 = DKK6.608; 1ECU = US$1.129; US$1 = DM1.735.
2. Danish turbines had a total share of over 50 per cent of the global wind turbine market in 1996. Therefore, in this chapter, Danish turbine cost figures are assumed to be representative of worldwide trends.
3. For normalised Danish wind conditions.
4. 'Ex works' means that no site work, foundation or grid connection costs are included. Ex works costs include the turbine as provided by the manufacturer, including the turbine itself, blades, tower and transport to the site.
5. Because output capacity (kW) changes in approximate proportion to swept area, a decline in $/m$^2$ cost is a rough indicator of a similar decline in $/kW.
6. Note: in terms of costs, only capital costs are reflected in this ratio. Any improvements in operation and maintenance (O&M) costs, equipment lifetime or equipment salvage values would not be reflected in the investment-per-production efficiency ratio.
7. Note: the improvement in $/kWh costs shown in Figure 4.4 (45% in 9–10 years) is slower than that suggested in Figure 4.2 (45% in 7 years). This is largely due to the fact that Figure 4.2 includes improvements in turbine siting, while Figure 4.4 represents wind energy costs under fixed siting conditions.
8. For operation and maintenance costs, the same profile (in relation to investment costs) is assumed as for land-based turbines, shown in Table 4.3.
9. EPRI (1997) suggests that wind turbines located in highly windy areas could achieve capacity factors of 40–45 per cent by the year 2005.
10. National assumptions on plant lifetime might be shorter, but calculations were adjusted to 40 years.
11. This may be significant when comparing conventional plants against dispersed small-scale wind turbines. Dispersed wind turbines often feed into the local grid near final consumers and thus have lower transmission and distribution losses.
12. Small-scale gen-sets are not designed for continuous operation and suffer from high maintenance needs under intensive operation. Gensets were therefore assumed to operate at full capacity for only four hours per day, based on the experience of local users.
13. The original analysis was conducted in 1981, and cost-effectiveness of all technologies is likely to have improved since that time. However,

wind pump technology is mature and is not advancing at the rate seen in larger wind turbines.

## Notes to Chapter 5

1. The propensity for IPP projects to be project-financed may be changing due to the cheaper financing terms often available through corporate finance and the cost reductions necessitated by increased competition facing developers in the generation market. See, for example, Jechoutek and Lamech (1995).
2. In reality, a capacity credit of 20–40 per cent is typically justified (see Chapter 3), but not always recognised by utilities.
3. This assumes a simple bank loan. Bonds can be traded on secondary markets, allowing the possibility for capital gains and losses as well; but such capital gains are also primarily determined by the interest rate rather than the company's profitability.
4. Senior debt ratio refers to the percentage of total finance provided by senior (not subordinated) debt.
5. Kahn (1995) and Wiser and Kahn (1996) illustrate that investors' ability to take advantage of the PTC requires greater use of high-cost equity, thus defeating much of the incentive effect which the tax credit was meant to provide for wind energy.
6. There exists substantial economic literature on the rationing of credit and the allocation of risks between creditors and equity owners. See, for example, Stiglitz and Weiss (1981), Easterbrook (1984) and Jensen and Meckling (1976).
7. The DSCR is not the only ratio considered by lenders. Others include the loan life coverage ratio and project life coverage ratio (see Mills and Taylor, 1994); but this current discussion focuses only on the DSCR which is considered particularly sensitive because of the annual nature of its constraints.
8. To facilitate comparisons, all other assumptions regarding capital costs, operating costs, capacity factor and so on were kept the same as in Wiser and Kahn's analysis of typical US conditions. Please see Wiser and Kahn (1996) for details of the cash flow model.
9. The less steep slope for the down-regulation curve is due to the existence of electro boilers in Norway and Sweden, which can be switched on and off at short notice to take advantage of low electricity costs.
10. Wind power's greater need for up-regulation power than conventional generators could nevertheless leave it vulnerable to short-term price spikes such as occurred in US electricity markets in late June 1998. However, such spikes do not occur instantaneously but rather tend to build over several days. Wind power plants should therefore be able to manage these risks through conservative bidding when such spikes are anticipated.

11. For a detailed review of the California green market, see Wiser and Pickle (1998).

## Notes to Chapter 6

1. Financial analysis always disregards externalities, since financial analysis is concerned only with those factors directly affecting project investors, as discussed in Chapter 5.
2. At the average 1997 exchange rate, 1ECU = US$1.129.
3. Breakdown of externality estimates of fossil fuels in Figure 6.4. Source: ExternE (1995), vol. 1, p. 163.

| *Not including global warming (in mECU/kWh)* | | *Global warming only (in mECU/kWh)* | *Total (in mECU/kWh)* |
|---|---|---|---|
| Coal | 6–16 | 10–18 | 16–34 |
| Oil | 11–12 | 6–12 | 17–24 |
| Gas | 0.7 | 4–8 | 5–9 |

Nuclear and hydro were provided as point estimates only. Note, the range of uncertainty on all of these values is extremely large. For more detailed information on how the numbers were derived, please refer to ExternE (1995).

4. Other helpful articles on the visual impact of wind turbines include, among others, Elliott (1994) and Wolsink (1989).

## Notes to the Epilogue

1. This figure includes the federal wind energy production tax credit. With no tax credit, the price would be approximately 0.7 cents/kWh higher.
2. The California Power Exchange also declared bankruptcy and ceased operations in early 2001, thereby eliminating a basic cornerstone of the restructured California market. The California Department of Water Resources replaced the investor-owned utilities as the electricity procurement agency for almost the entire state. By mid-2001, the California electricity market retained almost no resemblance to its original design of 1998.

# References


Abbott, I. H. and von Doenhoff, A. E. (1959), *Theory of Wind Sections, including a Summary of Airfoil Data* (New York: Dover).

Anayiotos, A. (1994), 'Infrastructure Investment Funds', *Public Policy for the Private Sector*, World Bank Private Sector Development Department, July.

Andersen, P. D. (1998), 'Financing Wind Farms: the Danish Experiences', Risø National Laboratory, Denmark, presented at Irish Energy Centre 'Wind Farms – The Banker's Perspective', Seminar, Dublin, April.

Andersen, P. D. and Hjuler Jensen, P. (1997), 'Wind Energy Technology in the 21st Century', in *European Wind Energy Conference Proceedings, EWEC '97*, Dublin, 6–9 October 1997, ed. R. Watson (Slane: Irish Wind Energy Association, 1998).

Anne Marie Simon Planning and Research (for British Wind Energy Association) (1996), *A Summary of Research Conducted into Attitudes to Wind Power from 1990 to 1996*, http://www.bwea.com/anmarie.htm, viewed 4 August 1997.

Arrow, K. J. and Lind, R.C. (1970), 'Uncertainty and the Evaluation of Public Investment Decisions', *American Economic Review*, **60**: 364–78, reprinted in R. Layard and S. Glaister (eds) (1996), *Cost–Benefit Analysis*, 2nd edn (Cambridge: Cambridge University Press).

AWEA (American Wind Energy Association) (1999), *Wind Energy is Fastest Growing Energy Source in World, Again*, AWEA news release, 7 January, http://www.awea.org/news/news990107.html, viewed 9 April.

AWEA (2001), *Wind Energy's Costs Hit New Low*, American Wind Energy Association, 6 March, http://www.awea.org/news/news010306cew.html, viewed 27 March 2001.

Bergey, M. (1998), *New Frontiers in Small Wind Turbine Development for Distributed Generation*, presented at World Energy Congress, Houston, 13–17 September, cited in *Wind Energy Weekly*, **17**, no. 819.

Betz, A. (1920), 'Das Maximum der Theoretisch Möglichen Ausnützung des Windes durch Windmotoren', *Zeitschrift für gesamte Turbinewesen*, **26**: 307.

Bhatia, R. and Pereira, A. (1988), *Socioeconomic Aspects of Renewable Energy Technologies* (Geneva: International Labour Office).

Björk, A. (1989), 'Airfoil Design for Variable RPM Horizontal Axis Wind Turbines', *Proceedings, EWEC '89*, European Wind Energy Conference and Exhibition, Glasgow.

Bond, G. and Carter, L. (1995), 'Financing Energy Projects: Experience of the International Finance Corporation', *Energy Policy*, **23**, no. 11.

Brealey, R. A., and Myers, S. C. (1996), *Principles of Corporate Finance*, 5th edn (London: McGraw-Hill).

Brown, M. H., and Yuen, M. (1994), 'Changing a Historical Perspective', *Independent Energy*, **24** (September) pp. 64–8.

BTM Consult and Danish Wind Turbine Manufacturers Association (1996), *Danish Wind Turbines Take 60 per cent of World Market*, press release, http://www.windpower.dk/news/p961231e.htm, viewed 4 August 1996.
BTM Consult (1997), *International Wind Energy Development: World Market Update 1996* (March).
BTM Consult (1998a), *International Wind Energy Development: World Market Update 1997* (March).
BTM Consult (1998b), *Ten Percent of the World's Electricity Consumption from Wind Energy*, http://www.btm.dk/Articles/fed-global/introduction.htm, viewed October 1998.
BTM Consult (1999), *International Wind Energy Development: World Market Update 1998* (March).
BTM Consult (2001), *International Wind Energy Development: World Market Update, 2000* (March).
BWEA (British Wind Energy Association) (1999), *The Economics of Wind Energy*, fact sheet,http://www.bwea.com/fs2econ.htm, viewed 4 February.
Byrne, J., Shen, B. and Wallace, W. (1998), 'The Economics of Sustainable Energy for Rural Development: a Study of Renewable Energy in Rural China', *Energy Policy*, **26**, no. 1.
CADDET (1997), 'A Second Generation Wind/Diesel System', *Renewable Energy Newsletter*, IEA/OECD (March).
CADDET (1998a), 'Wind Turbines Help New Zealand Farmers to Pump Water', *Renewable Energy Project*, IEA/OECD, http://194.178.172.86/register/data-re/ccr01798.htm, viewed 15 January.
CADDET (1998b), 'Collaborative Project Provides Wind-Diesel Hybrids to Northern Territories of Russia', *Renewable Energy Project*, IEA/OECD, http://194.178.172.86/register/data-re/ccr01932.htm, viewed 15 January.
CADDET (1998c), 'The Terschelling Photovoltaic and Wind Energy Project', *Renewable Energy Project*, IEA/OECD, http://194.178.172.86/register/data-re/ccr01769.htm, viewed 15 January.
CADDET (1998d), *Renewable Energy Newsletter*, IEA/OECD, issue 1/98 (February).
California Assembly Bill No. 1890, Chapter 854 (1996), signed 23 September.
California Assembly Bill No. 29 (2001), signed 11 April.
California Energy Markets (2000), No. 597, 15 December, p. 3, Energy NewsData Corp.
California Energy Markets (2001), No. 614, 20 April, pp. 15–17, Energy NewsData Corp.
California Senate Bill, No. 5 (2001), signed 11 April.
CEC (California Energy Commission) (1997), *Policy Report on AB 1890 Renewables Funding, Report to the Legislature*, P500-97-002 (March).
Chew, W. (1995a), 'Private Power Project Rating is Critical for Permanent Financing', *Private Power Executive* (March–April).
Chew, W. (1995b), 'Key Contract Conditions Critical for Investment-Grade Ratings', *Private Power Executive* (May–June).
Chew, W. (1995c), 'Power Project Ratings Keyed to Production, Cash Flows', *Private Power Executive* (July–August).

Cox, A. J., Blumenstein, C. J. and Gilbert, R. J. (1991), 'Wind Power in California: a Case Study of Targeted Tax Subsidies', in Richard J. Gilbert (ed.), *Regulatory Choices: A Perspective on Developments in Energy Policy* (Berkeley, CA: University of California Press).

CPUC (California Public Utilities Commission) (1993), *California's Electric Services Industry: Perspectives on the Past, Strategies for the Future*, Division of Strategic Planning (February).

CPUC (California Public Utilities Commission) (1997), *Interim Opinion on Public Purpose Programs: Threshold Issues*, Decision 97-02-014, 5 February.

CPUC (California Public Utilities Commission) (2001), *Interim Opinion: Implementation of Public Utilities, Code Section 399. 15(b), Paragraphs 4–7; Load Control and Distributed Generation Initiatives*, Desicion 01-03-073, 27 March.

Danish Energy Agency (1997), *Havmølle-handlingsplan for de danske farvande (Plan of Action for Offshore Windfarms in Danish Waters)*, Elselskabernes og Energistyrelsens Arbejdsgruppe for havmøller (June).

Danish Wind Turbine Manufacturers Association (1998a), *Wind Turbine Markets*, www.windpower.dk/stat/tab19.htm, viewed October 1998.

Danish Wind Turbine Manufacturers Association (1998b), *Wind Power Note*, no. 19 (May).

Danish Wind Turbine Manufacturers Association (1999), *Wind Power Note*, no. 22 (April).

Diaz, H. F., Wolter, K. and Woodruff, S. D. (1992), *Proceedings: International COADS Workshop*, (Boulder, CO), January 1992.

Easterbrook, F. H. (1984), 'Two Agency-Cost Explanations of Dividends', *American Economic Review*, **74**, no. 4 (September), pp. 650–9, reprinted in C. W. Smith Jr. (ed.) (1990), *The Modern Theory of Corporate Finance*, 2nd edn (London: McGraw-Hill).

EDF (Environmental Defense Fund) (1999), *Comparing 'Green' Electricity Products in California*, http://www.edf.org/programs/Energy/green power/ c_providers.html, viewed October 26, 1999.

EEA (European Environment Agency) (1996), *Environmental Taxes: Implementation and Environmental Effectiveness*, Environmental Issues Series no. 1 (August).

EIA (Energy Information Administration) (1992), *Federal Energy Subsidies: Direct and Indirect Interventions in Energy Markets*, SR/EMEU/92-02, Distribution Category UC-98, US Department of Energy, Washington, DC.

Elliott, D. A. (1994), 'Public Reactions to Windfarms: the Dynamics of Opinion Formation', *Energy & Environment*, **5**, issue 4.

Ellison, C. T., Brown, A. B. and Rader, N. (1997), *Scoping Paper on Restructuring-Related System Operation and Transmission Issues*, for the National Wind Coordinating Committee, 24 January.

Elsamprojekt A/S (1997), *Vindmøllefundamenter i havet* (Foundations for Offshore Wind Turbines) (Fredericia).

Emeis, S. and Frandsen, S. (1993), 'Reduction of Horizontal Wind Speed in a Boundary Layer with Obstacles', *Boundary Layer Meteorology*, **64**: 297–305.

Energistyrelsen (The Danish Energy Agency) (1994), *Privatejede Vindmøllers Økonomi* (The Economics of Privately Owned Wind Turbines).

Energy 21 (The Danish Government's Action Plan for Energy) (1997), http://www.ens.dk/e21/e21uk/Underkap/23.htm, viewed October 1997.

EPRI (Electric Power Research Institute) (1997), *Renewable Energy Technology Characterizations*, TR-109496, December, cited by Izaak Walton League and Minnesotans for an Energy-Efficient Economy, http://www.me3.org/issues /wind/iwlanspirp.html, viewed 17 February 1999.

Eurec-Agency (1996), *The Future for Renewable Energy* (London: James & James).

European Commission (1997), *Energy for the Future: Renewable Sources of Energy*, White Paper for a Community Strategy and Action Plan, COM(97) 599, Communication from the Commission, http://www.europa.eu.int/ en/comm/dg17/599fi_en.htm, viewed 26 November 1997.

ExternE (1995), *Externalities of Energy*, European Commission Directorate-General XII: Science, Research and Development, EUR 16520 EN, Office for Official Publications of the European Communities, L-2985, Luxembourg.

Farhar, B. C. and Houston, A. H. (1996), *Willingness to Pay for Electricity from Renewable Energy*, National Renewable Energy Laboratory, NREL/TP-460-21216 (September).

Fenhann, J., Morthorst, P. E., Schleisner, L., Møller, F. and Winther, M. (1998), *Samfundsøkonomiske omkostninger ved reduktion af drivhusgas emissioner* (National Economic Costs by Reducing Greenhouse Gas Emissions, in Danish) (Copenhagen: The Danish Ministry of Environment and Energy).

FERC (US Federal Energy Regulatory Commission) (2001), Docket No. EL01-64, California Cogeneration Council et al., filed 5 April.

Francis, J. C. (1993), *Management of Investments*, 3rd edn (London: McGraw-Hill).

Frandsen, S. and Christensen, C. J. (1992), 'Accuracy of Estimation of Energy Production from Wind Power Plants', *Wind Engineering*, **16**, no. 5, pp. 257–68.

Fuglsang, P. and Aagaard Madsen, H. (1996), 'Numerical Optimization of Wind Turbine Rotors', in *European Union Wind Energy Conference, Proceedings, EWEC '96*, Göteborg, 20–24 May, ed. A. Zervis, H. Ehmann and P. Helm (Bedford: H. S. Stephens & Associates) pp. 679–82.

Gerdes, G. J., Santer F. and Klosse, R. (1997), 'Overview and Development of Procedures on Power Quality Measurements of Wind Turbines', in *European Wind Energy Conference Proceedings, EWEC '96*, Dublin, 6–9 October 1997, ed. R. Watson (Slane: Irish Wind Energy Association, 1998).

Gilbert, R. J. (1991), 'Issues in Public Utility Regulation', in *Regulatory Choices: A Perspective on Developments in Energy Policy*, ed. Richard J. Gilbert (Berkeley, CA: University of California Press).

Gipe, P. (1993), *Wind Power for Home & Business* (London: Chelsea Green).

Gipe, P. (1995), *Wind Energy Comes of Age* (New York: John Wiley).

Gipe, P. (1997), *Tilting at Windmills: Public Opinion toward Wind Energy*, gopher://gopher.igc.apc.org/0/orgs/awea/faq/surv/gipe, viewed 27 September, 1997.

Glauert, H. (1935), 'Airplane Propellers', in W. F. Durand, *Aerodynamic Theory* (New York: Dover, 1963).
Golding, E. W. (1976), *The Generation of Electricity by Wind Power*, 2nd edition (New York: Halsted Press).
Grubb, M. J. and Meyer, N. I. (1993), 'Wind Energy: Resources, Systems, and Regional Strategies', in T. B. Johansson, H. Kelly, A. K. N. Reddy and R. H. Williams (eds), *Renewable Energy: Sources for Fuels and Electricity* (Washington, DC: Island Press, pp. 157–212.
Hakimian, H. and Kula, E. (1995), 'The Environment and Project Appraisal', *Investment and Project Appraisal*, University of London, Centre for International Education in Economics.
Hall, D. A. and Blowes, J. H. (eds) (1995), *Working Cost and Operational Report: Stationary Engines and Gas Turbines*, Transactions of the Institution of Diesel and Gas Turbine Engineers, Bedford, 1994.
Hammarlund, K. (1996), 'Improving Acceptance in Wind Power Planning', *Proceedings of 1996 European Union Wind Energy Conference*, Göteborg, May.
Hamrin, J. G., and Rader, N. (1992), 'Non-Utility Power Development in the USA: the Independent Generators', *Energy Policy* (November).
Hansen, J. C., Paulsen, U. S., El Hewehy, A. and Mansour, E. S. (1997), 'Hurghada Wind Energy Technology Center: Background, Facilities and Perspectives', *Environment '97 Conference*, Cairo, February.
Hasson, G., Redlinger, R. and Bouille, D. (1998), *Implications of Electric Power Sector Restructuring on Climate Change Mitigation in Argentina: Economics of GHG Limitations*, Instituto de Economía Energética and UNEP Collaborating Centre on Energy and Environment, Denmark, September.
Heifele, W. et al. (1981), *Energy in a Finite World* (Cambridge, Mass.: International Institute of Applied System Analysis/Ballinger).
Hopkins, W. (1999), 'Small to Medium Size Wind Turbines: Local Use of a Local Resource', *Renewable Energy*, **16**: 944-7
Hoppe-Kilpper, M., Kleinkauf, W., Schmid, J., Stump N. and Windheim, R. (1996), 'Experiences with Over 1000 MW Wind Power Installed in Germany', *Proceedings of 1996 European Union Wind Energy Conference*, Göteborg, May.
Hunt, S., and Shuttleworth, G. (1996), *Competition and Choice in Electricity* (New York: John Wiley).
IEA (International Energy Agency) (1996a), *Energy Policies of IEA Countries: Sweden 1996 Review*, IEA/OECD.
IEA (International Energy Agency) (1996b), *IEA R&D Wind Annual Report 1996*.
IEA (International Energy Agency) (1997a), R&D WTS Annex XV, *Review of Progress in the Implementation of Wind Energy by the Member Countries of the IEA During 1996*, July 1997.
IEA (International Energy Agency) (1997b), *Renewable Energy Policy in IEA Countries*, vol. 1: *Overview*, IEA/OECD.
IEA (International Energy Agency) (1998), *IEA Wind Energy Annual Report 1997*, National Renewable Energy Laboratory, Golden, CO, September 1998.

IREDA (Indian Renewable Energy Development Agency) (1997), *Renewable Energy Financing Guidelines*.

IWLA (1999), 'NSP Should Acquire an Additional 400 MW of Wind Power', filing by Izaak Walton League of America – Midwest Office to Minnesota Public Utilities Commission, http://www.me3.org/issues/wind/iwlanspirp.html, viewed 17 February 1999.

Jechoutek, K. G. and Lamech, R. (1995), 'Private Power Financing: from Project Finance to Corporate Finance', *Public Policy for the Private Sector*, World Bank Industry and Energy Department (October).

Jenkins, G. (1996), 'Survey of Employment in the UK Wind Energy Industry', Wind Energy Conversion 1996, *Proceedings of the 18th British Wind Energy Association Conference*, September.

Jensen, M. C. and Meckling, W. H. (1976), 'Theory of the Firm: Managerial Behavior, Agency Costs, and Ownership Structure', *Journal of Financial Economics*, **3**, no. 4 (October) pp. 305–60, reprinted in C. W. Smith Jr. (ed.) (1990), *The Modern Theory of Corporate Finance*, 2nd edn (London: McGraw-Hill).

Kahn, E. (1995), *Comparison of Financing Costs for Wind Turbine and Fossil Powerplants*, Lawrence Berkeley Laboratory, LBL-36122, UC-1320 (February).

Kahn, E. (1996), 'The Production Tax Credit for Wind Turbine Powerplants is an Ineffective Incentive', *Energy Policy*, **24**, no. 5.

Kahn, E. P., Meal, M., Doerrer, S. and Morse, S. (1992), *Analysis of Debt Leveraging in Private Power Projects*, Lawrence Berkeley Laboratory, LBL-32487, UC-400 (August).

Kula, E. (1987), 'Social Interest Rate for Public Sector Appraisal in the United Kingdom, the United States and Canada', *Project Appraisal*, **2**, no. 3 (September), pp. 169–75.

Kwant, K.W., Novem (the Netherlands Agency for Energy and the Environment) (1996), 'Fiscal Support for Renewables in the Netherlands', *CADDET Renewable Energy Newsletter*, IEA/OECD, Issue 3/96 (September).

Landberg, L. (1998), 'Predicting the Power Output from Wind Farms', in *European Wind Energy Conference, Proceedings, EWEC '97*, Dublin, 6–9 October 1997, ed. R. Watson (Slane: Irish Wind Energy Association, 1998) pp. 747–54.

Landberg, L. (1999), 'Operational Results from a Physical Power Prediction Model', presented at the *1999 European Wind Energy Conference and Exhibition*, 1–5 March, Nice, France.

Landberg L. and Watson, S. J. (1994), 'Short-Term Prediction of Local Wind Conditions', *Boundary-Layer Meteorology*, **70**: 171–95.

Layard, R. and Glaister, S. (1996), *Cost-Benefit Analysis*, 2nd edn (Cambridge: Cambridge University Press).

Li Junfeng (1996), *Renewable Energy Development in China*, Center for Renewable Energy Development: Energy Research Institute of State Planning Commission, Workshop for the Assessment of GHG Mitigation, 12–15 November 1996, Beijing.

Lindley, David (1996), 'A Study of the Integration of Wind Energy into the National Energy Systems of Denmark, Wales and Germany as Illustrations of Success Stories for Renewable Energy', *Proceedings of 1996 European Union Wind Energy Conference*, Göteborg, May.

MacRae, A. N. and Saluja, G. S. (1989), 'Wind Turbine Performance and Reliability: a Risk Based Analysis', *Proceedings: Windpower '89* (AWEA), San Francisco, California, 24–7 September.

ME3 (1998a), 'Is Wind Power NSP's Least-Cost Resource?', *Sustainable Minnesota*, **8**, no 3, Minnesotans for an Energy-Efficient Economy, http://www.me3.org/newsletters/me3sum98.html, viewed 17 February 1999.

ME3 (1998b), 'DPS Confirms Wind Energy is NSP's Least-Cost Resource', *Sustainable Minnesota*, **8**, no 4, Minnesotans for an Energy-Efficient Economy, http://www.me3.org/newsletters/me3fal98.html, viewed 17 February 1999.

ME3 (1999), 'Minnesota Public Utilities Commission Votes 4–0 Ordering NSP to Build Additional 400 MW of Wind Power', *Sustainable Minnesota Energy News*, 22 January, Minnesotans for an Energy-Efficient Economy, http://www.me3.org/news/012299ns.html, viewed 17 February 1999.

Mead, W. J. and Denning, M. (1991), 'Estimating Costs of Alternative Electric Power Sources for California', in *Regulatory Choices: A Perspective on Developments in Energy Policy*, ed. Richard J. Gilbert (Bekerly, CA: University of California Press,).

Milborrow, D. J. and de Mouy, L. D. (1989), 'Survey of Energy Resources', report from World Energy Council.

Mills, S. J. and Taylor, M. (1994), 'Project Finance for Renewable Energy', *Renewable Energy*, **5**, part I, pp. 700–8.

Minister of Economic Affairs, Netherlands (1996), *Current Lines Towards an Electricity Market: The Framework for a New Electricity Act in the Netherlands* (Ministry of Economic Affairs, Netherlands).

Mitchell, C. (1995), 'The Renewables NFFO', *Energy Policy*, **23**, no. 12.

Mitchell, C. (1996), 'Future Support of Renewable Energy in the UK – Options and Merits', *Energy & Environment*, **7**, no. 3.

Mitchell, C. (1997), 'The Renewable Non-Fossil Fuel Obligation: the Diffusion of Technology by Regulation', *TIP Workshop on Regulation and Innovative Activities*, Vienna (February).

Modigliani, F. and Miller, M. H. (1958), 'The Cost of Capital, Corporation Finance, and the Theory of Investment', *American Economic Review*, **48** (June) pp. 261-97.

Mortensen, N. G. and Said, Usama (1996), *Wind Atlas for the Gulf of Suez: Measurements and Modelling, 1991–95*, Risø National Laboratory, Denmark and NREA, Egypt.

Morthorst, P. E. (1996), 'Independent Power Production in Denmark: Wind Turbines', *ENER Bulletin 18.96*, European Network for Energy Economics Research.

Munksgaard, J., Larsen, A., Pedersen, M. R., Pedersen, J. R., Jordal-Jørgensen, J., Jensen, T., Karnøe, P., and Jørgensen, U. (1996) *Social Assessment of*

*Wind Power,* AKF Institute of Local Government Studies, Denmark, http://www.akf.dk/eng/wind.htm, viewed 7 August, 1997.

Netherlands Ministry of Economic Affairs (1997), *Renewable Energy: Advancing Power: Action Programme for 1997–2000.*

New Energy (1998), no. 1 (December), p. 54, Bundesverband WindEnergie (BWE e.V), Osnabrück.

Nielsen, L. H. (ed.) (1994), *Vedvarende energi i stor skala til el og varmeproduktion* ('Renewable Energy for Large-Scale Power and Heat Production in the Future Danish Energy System', Main Report), Risø National Laboratory (December).

Nielsen, P. (1997), *Mini-Analyse for Vindmøller* (Small Analysis for Wind Turbines), Aalborg: Energy and Environmental Data – EMD).

Nielsen, L. H., and Morthorst, P. E. (eds) (1998), *Fluktuerende vedvarende energi i el- og varmeforsyningen – det mellemlange sigt* (System Integration of Wind Power under Liberalised Electricity Market Conditions, Medium-term Aspects), Risø National Laboratory (April).

Novem (The Netherlands Agency for Energy and the Environment) (1998), *Green Financing,* http://www.novem.org/netherl/green.htm, viewed 28 January, 1998.

O'Gallachoir, B. (1998), Irish Renewable Energy Information Office, personal communication, 10 July.

OECD/IEA (OECD Nuclear Energy Agency, and International Energy Agency) (1998), *Projected Costs of Generating Electricity – Update 1998*, Paris.

Øye, S. (1992), *FLEX4: Computer Code for Wind Turbine Load Simulation* (Technical University of Denmark).

Pedersen, A. M. J., Jensen, L. E. and Jørgensen, U. K. (1994), 'Reliability of Wind Farm Production Forecasts', *European Wind Energy Conference Proceedings: EWEC '94*, Thessaloniki, Greece.

Petersen, E. L., Mortensen, N. G. and Landberg, L. (1994), 'Wind Resource Estimation and Siting of Wind Turbines', *European Directory of Renewable Energy Suppliers and Services*, pp. 181–190.

Petersen, E. L., Mortensen, N. G., Landberg, L., Højstrup, J. and Frank, H. P. (1998a), 'Wind Power Meteorology, Part I: Climate and Turbulence', *Wind Energy,* **1**: 2–22.

Petersen, E. L., Mortensen, N. G., Landberg, L., Højstrup, J. and Frank, H. P. (1998b), 'Wind Power Meteorology, Part II: Siting and Models', *Wind Energy,* **1**: 55–72.

*Private Power Executive* (1997), 'Before Seeking Project Financing, Examine All Sources', May–June, pp. 25–35.

Rader, N. A. (1996), American Wind Energy Association, 'The Renewables Portfolio Standard in California as Envisioned by the American Wind Energy Association', http://www.igc.apc.org/awea/pol/rpsca.html, viewed 7 May.

Rader, N. A. (1997), 'Information on Status of Renewables Portfolio Standard', personal communication, 20 August.

Rader, N. A. (1998), *Green Buyers Beware: A Critical Review of 'Green Electricity' Products* (Public Citizen, Washington, DC, October).

Rader, N. A, and Short, W. P. (1998), 'Competitive Retail Markets: Tenuous Ground for Renewable Energy', *The Electricity Journal*, **11** no. 3.
Razavi, H. (1996), *Financing Energy Projects in Emerging Economies* (Tulsa, OK: PennWell Publishing).
RCG/Hagler Bailly (1994), *New York State Environmental Externalities Cost Study*, Empire State Electric Energy Research Corporation, and New York State Energy Research and Development Authority.
REPSource (1998), 'Global Financiers of Renewable Energy Enterprises', *Newsletter of the International Network of Renewable Energy Project Support Offices*, Winrock International, **3**, no. 1.
Righter, R. (1996), *Wind Energy in America* (University of Oklahoma Press).
Sarkar, S. K. and Bhatia, D. (1997), *Indian Wind Energy Scenario* (Indian Renewable Energy Development Agency).
Schleisner, L. and Nielsen, P. S. (1997), *External Costs Related to Power Production Technologies: ExternE National Implementation for Denmark*, Risø National Laboratory, December.
Sharma, R. A., McGregor, M. J. and Blyth, J. F. (1991), 'The Social Discount Rate for Land Use Projects in India', *Journal of Agricultural Economics*, **42**: 86–93.
Shephard, D. G. (1990), *Historical Development of the Windmill*, NASA Contractor Report 4337, DOE/NASA-5266-2.
Skytte, K. (1997), 'Fluctuating Renewable Energy on the Power Exchange', Risø National Laboratory, Denmark, discussion paper, presented at *Energy Economics Conference, University of Warwick*, 8–9 December.
Skytte, K. (1999), 'The Regulating Power Market on the Nordic Power Exchange Nord Pool: An Econometric Analysis', *Energy Economics*, **21**.
Stiglitz, J. E. and Weiss, A. (1981), 'Credit Rationing in Markets with Imperfect Information', *American Economic Review*, no. 71, pp. 393–410.
Stoft, S.,Webber C. and Wiser, R. (1997), *Transmission Pricing and Renewables: Issues, Options, and Recommendations*, Lawrence Berkeley National Laboratory, LBNL-39845, UC-1321 (May).
Suresh, R. (1997), Tata Energy Research Institute, personal communication, 27–29 August.
Suresh, R. (1999), Tata Energy Research Institute, personal communication, 18 March.
Swisher, J. N., Jannuzzi, G. M. and Redlinger, R. Y. (1997), *Tools and Methods for Integrated Resource Planning: Improving Energy Efficiency and Protecting the Environment*, UNEP Collaborating Centre on Energy and Environment, Risø National Laboratory, Denmark.
Tande, J. O. and Hansen, J. C. (1996), 'Wind Power Fluctuation's Impact on Capacity Credit', *Proceedings of 1996 European Union Wind Energy Conference*, Göteborg, May.
Tande, J. O., Nørgård, P., Sørensen, P., Søndergård, L., Jørgensen, P., Vikkelsø, A., Dyring Kledal, J. and Christensen, J. S. (1996), *Power Quality and Grid Connection of Wind Turbines*, Summary Report, Risø National Laboratory, Roskilde, Denmark, Risø-R-853.
Tangler, J. L. and Somers, D. M. (1985), 'Advanced Airfoil for HAWTS', *Proceedings Windpower '85* (San Francisco, CA).

Troen, I. and Petersen E. L. (1989), *European Wind Atlas*, Risø National Laboratory, Roskilde, Denmark.

Turkson, J. K. (ed.) (2000), *Power Sector Reform: Process and Implementation Experiences in Sub-Saharan Africa* (Basingstore: Macmillan – now Palgrave).

UN (1997), *Kyoto Protocol to the United Nations Framework Convention on Climate Change*, Conference of the Parties, Third Session, Kyoto, 1–10 December.

UNEP (1998), *The Economics of Greenhouse Gas Limitations: Technical Guidelines*, Draft, February, UNEP Collaborating Centre on Energy and Environment, Risø National Laboratory, Roskilde, Denmark.

United Nations (1964), *New Sources of Energy*, Proceedings of the Conference, Rome, 21–31 August 1961, vol. 7: *Wind Power* (New York: United Nations).

US Department of Energy (1985), Five Year Research Plan 1985–1990, *Wind Energy Technology: Generating Power from the Wind*, DOE/CD-T11.

USEIA (US Energy Information Administration) (1995), *US Wind Energy Potential: The Effect of the Proximity of Wind Resources to Transmission Lines*, cited in *Private Power Executive*, 'Wind Energy Potential Exceeds All Existing US Power Generation', September–October 1995.

USEIA (US Energy Information Administration) (1996), *Emissions of Greenhouse Gases in the United States 1995*, US Department of Energy, DOE/EIA-0573(95), October, http://www.eia.doe.gov/oiaf/1605/gg96rpt, viewed 23 December 1998.

USEPA (US Environmental Protection Agency) (1998), AIRSData, EPA Office of Air Quality Planning And Standards, Source SIC Report, United States Air Pollution Sources, http://www.epa.gov/airsdata/srcsic.htm, viewed 23 December 1998.

van Wijk, A. J. M., Coelingh, J. P. and Turkenburg, W. C. (1993), 'Global Potential for Wind Energy', *Proceedings of the European Wind Energy Conference, 1991* (EWEC-'91), Amsterdam, Netherlands.

Vindmølleindustrien (1996), 'Employment in the Wind Power Industry', *Wind Power Note*, Danish Wind Turbine Manufacturers Association (March).

Vindmølleindustrien (1997), *Statistics: Installed Wind Power Capacity in Denmark in MW*, Table 12, Danish Wind Turbine Manufacturers Association, http://www.windpower.dk/stat/tab12.htm, viewed 28 August 1997.

WEC (World Energy Council) (1994), *New Renewable Energy Resources – A Guide to the Future* (London: Kogan Page).

Wilson, R. E. and Lissaman, P. B. S. (1974), *Applied Aerodynamics of Wind Power Machines*, Report NSF/RA/N-74113, Oregon State University.

Wind Energy Weekly (1996–1997), American Wind Energy Association, vols 15–16, nos 725, 736, 745, 746.

Wind Energy Weekly (1997), 'Ancillary Services Pricing Poses Questions for Wind', American Wind Energy Association, vol. 16, no. 755 (14 July).

Wind Energy Weekly (1998), American Wind Energy Association, vol. 17, no. 816 (28 September).

Wind Energy Weekly (1999), American Wind Energy Association, vol. 18, no. 876 (9 December).

Windpower Monthly (1997), 'New Year Green Credits Legislation: Wind Project First Beneficiary', and 'Liberalisation Law Puts Cap on Wind Output', vol. 13, no. 12.
Windpower Monthly (1998a), 'Cash Plentiful for Projects in Developing Countries' (January).
Windpower Monthly (1998b), 'Price of Wind Drops to Fully Competitive Level' (May).
Windpower Monthly (1998c), 'Renewables and the Real World', and 'Power Pool Supplies Storage for Wind' (April).
Windpower Monthly (1998d), 'Massachusetts goes for Renewables: Deregulation Law Includes RPS', and 'Deregulation in Denmark: Protection for Wind Stays in Place', and 'Wind Wire: Germany: Barrier Ahead', and 'Green Generators to Pool Output', vol. 14, no. 1(January).
Windpower Monthly (1998e), 'First Day Excitement and Foreboding', and 'Trading Surprise', and 'More Trouble in India: Tax Evasion Scam Linked to Wind', and 'New Energy Law Kicks In: More Subsidies Breed Hope', vol. 14, no. 3 (March).
Windpower Monthly (1998f), 'Steps to Remove the Bureaucratic Barriers', vol. 14, no. 4 (April).
Windpower Monthly (1998g), 'India Potential Undiminished: Changes for the Better on the Way', vol. 14, no. 5 (May).
Wiser, R. and Kahn, E. (1996), *Alternative Windpower Ownership Structures: Financing Terms and Project Costs*, Lawrence Berkeley National Laboratory, LBNL-38921, UC-1321 (May).
Wiser, R., and Pickle, S. (1997a), *Financing Investments in Renewable Energy: The Role of Policy Design and Restructuring*, Lawrence Berkeley National Laboratory, LBNL-39826, UC-1321 (March).
Wiser, R., and Pickle, S. (1997b), *Green Marketing, Renewables, and Free Riders: Increasing Customer Demand for a Public Good*, Lawrence Berkeley National Laboratory, LBNL-40532, UC-1321 (September).
Wiser, R. and Pickle, S. (1998), *Selling Green Power in California: Product, Industry, and Market Trends*, Lawrence Berkeley National Laboratory, LBNL-41807 (May).
Wolsink, M. (1989), 'Attitudes and Expectancies about Wind Turbines and Wind Farms', *Wind Engineering*, **13**, no. 4.
Wolsink, M. (1996), 'Dutch Wind Power Policy: Stagnating Implementation of Renewables', *Energy Policy*, **24**, no. 12.
Wolsink, M. (1998), personal communication, 30 November.
World Bank (1996), *Rural Energy and Development: Improving Energy Supplies for Two Billion People* (Washington, DC).
Worldwatch (1999), *World Nuclear Power Electrical Generating Capacity*, http://www.worldwatch.org/alerts/990304a.html, viewed 9 April.
Zerbe, R. O. and Dively, D. D. (1994), *Benefit–Cost Analysis: In Theory and Practice* (HarperCollins).

# Index